WITHDRAWN

ALSO BY THE SAME AUTHOR

No. BP32	How to Build Your Own Metal and Treasure Locators
No. BP35	Handbook of IC Audio Pre-amplifier and Power Amplifier Construction
No. BP37	*50 Projects Using Relays, SCR's and Triacs*
No. BP39	50 (FET) Field Effect Transistor Projects
No. BP43	How to Make Walkie-talkies
No. BP48	Electronic Projects for Beginners
No. BP57	How to Build Your Own Solid State Oscilloscope
No. BP67	Counter Driver and Numeral Display Projects
No. BP75	Electronic Test Equipment Construction
No. BP79	Radio Control for Beginners
No. BP84	Digital IC Projects
No. BP90	Audio Projects
No. BP93	Electronic Timer Projects

IC PROJECTS FOR BEGINNERS

by
F. G. Rayer
T.Eng., (CEI), Assoc.IERE

BERNARD BABANI (publishing) LTD
THE GRAMPIANS
SHEPHERDS BUSH ROAD
LONDON W6 7NF
ENGLAND

Although every care has been taken with the preparation of this book, the publishers or author will not be held responsible in any way for any errors that might occur.

First Published – April 1982

British Library Cataloguing in Publication Data
Rayer, F. G.
IC projects for beginners – (BP97)
1. Integrated circuits – Amateurs' manuals
I. Title
621.381'37 TK9965

ISBN 0 85934 072 4

Printed and bound in Great Britain by Cox & Wyman Ltd, Reading

CONTENTS

Page

CHAPTER 1

COMPONENTS

Integrated Circuits

An IC consists of a small moulded package in which a number of transistors, diodes and other items are incorporated. Semiconductor junctions and conductors are formed in the device during manufacture, and once preparations for a particular IC are completed, great numbers can be produced rapidly and at relatively low cost.

Component density in the IC is very high. A large number of transistors and other items can be included in a package which with some ICs is no larger than a single transistor of ordinary type. The IC thus allows simplification in building – internal interconnections for devices are already present. Improved results are also usual, as the IC often has complicated circuitry which it would be tiresome to duplicate with separate transistors and other components.

Because there is such a wide range of ICs, the type and size of package varies enormously. Some resemble a small transistor, with only three leads. Many have about eight to sixteen connections or pins, though other large ICs have many more than this.

Figure 1 shows some ICs, and a dual-in-line 8-pin holder at A. The DIL type of holder is also made for 14-pin and other ICs. B shows ICs able to fit in holder A. C is pin numbering, when seen from above. Pin 1 end is shown by a dot, 1, dent, or similar means. As most ICs could be fitted either way round, care must obviously be taken that an IC is not reversed. Some holders have similar marks, so that the way the IC fits is identified.

D is the 14-pin IC, and 16-pin ICs are similar. Some have heat sink tabs, or staggered pins.

When first inserting an IC into its holder, some care should be taken. The pins are often not correctly spaced to slip into the holder sockets. If wide, the IC can usually be pressed sideways on a level surface, to correct this. Note that each pin

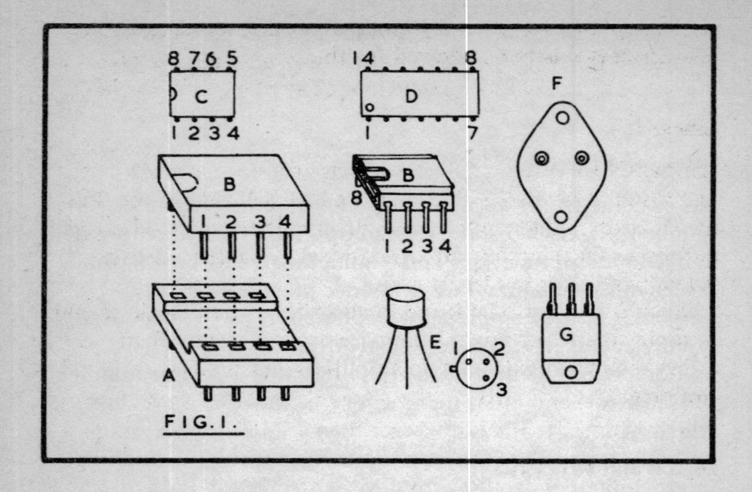

FIG. 1.

goes properly into its socket, so that the IC lies flat on the holder. To remove the IC, if this is ever necessary, gently lever it up a little at a time from each end.

It is of some advantage to use holders, as the ICs can be changed or removed easily. If an IC is soldered directly into place, lengthy heating should be avoided.

E is a small can type IC, with wire leads. F and G are power supply ICs, used for voltage control, and to be attached to heat sinks. They can contain all the necessary circuitry for voltage regulation, automatic overload protection, protection against over-heating, and against short-circuits. Quite a complicated circuit would be needed to produce the same result with separate transistors and other items.

Peripheral Components

The integrated circuit generally operates in conjunction with some external components. Capacitors and resistors are often fitted in this way. The number of external components depends on the IC, and the way in which it is used. The same IC may be used in various ways, some needing more external components. Values may also be modified, to suit the purpose.

These items are generally grouped on the board, near the IC. Space needs to be allowed for this.

Linear ICs

Linear ICs are often amplifiers which introduce as little distortion as possible. They may be high gain, low noise, low power types, intended for audio preamplifiers and similar purposes. Or they may incorporate devices able to deliver several watts, for loudspeaker use.

Such ICs can also be fitted in all sorts of circuits where one or more transistors might otherwise be employed. Their utility is not limited to audio amplifiers and similar equipment.

Digital and Other ICs

The digital IC is not used for amplification in the usual way, but can provide various on-off and similar functions, such as are associated with logic gates, decade counters and decoders, and so on. There is a great range of these, for numerous functions.

ICs also find useful applications in radio frequency and intermediate frequency amplifiers, oscillators, power supply regulators, and in radio receivers and electronic equipment generally.

Many of the popular ICs have alternatives of different manufacture and coding. But caution should be exercised in any substitution of types listed. Usually, a near equivalent is not suitable, and may introduce all sorts of obscure difficulties.

Unless means of testing are to hand, the use of surplus and untested ICs is best avoided. These may have a high failure rate and cause needless disappointment. It is simply not worth while fitting them.

Boards

These are readily available with perforations based on 0.1in and 0.15in grids, Figure 2. The required size of project for a board can be seen by noting the number of rows of holes,

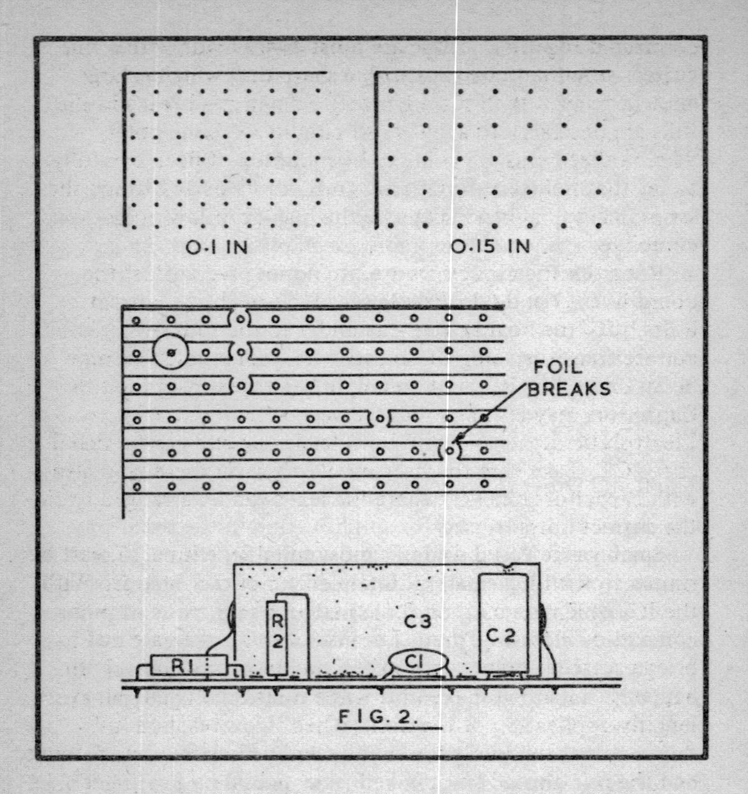

FIG. 2.

allowing a little extra for fixing holes, or any unusually large components which will be used. When the foil strips are to be used as conductors, the board must be arranged with these running in the correct direction. Large boards are readily sawn for small projects by means of a small toothed saw.

With some projects, either 0.1in or 0.15in board could be used, as those integrated circuits having wire leads can be arranged to fit the perforations. But the DIL type of IC and its holder requires the 0.1in board. The 0.15in board gives a little extra space for both soldering and components.

Interruptions to the foil conductors are often needed, to avoid short circuits between items which must not be connected together. These are most easily made with a foil cutter, or with a few turns with a sharp drill which is large enough to span the foils, Figure 2. Sometimes breaks in the foils are necessary to avoid short circuits at fixing holes, or where other items are going to be mounted. Check carefully to see that points of foil do not turn outwards and touch the strips nearby, and to make sure that each break is in fact complete.

Resistors and capacitors are often mounted against the board like R1 and C1. Pass the leads through the correct holes, turn the board over, and solder to the foils or connections, then snip off excess wire. Some resistors may be mounted like R2, with one wire bent over from the top. Capacitors may also be mounted vertically in this way. Electrolytic capacitors with both leads one end can be fitted as for C2. Take care to observe polarity with these, and also with capacitors like C3, where the leads can be arranged to fit the correct holes.

Small gauge cored solder is most suitable, with a 15 watt or similar iron with a small bit intended for circuit boards. With the ICs which are used on 0.1in matrix board, rows of pins come quite close together. Because of this, take care not to bridge adjacent points with excess amounts of solder. If this happens, it may be shaken off while molten. In general, avoid lengthy application of the iron. When it has reached full temperature, the joints should be seen to flow in a few seconds, and the iron should be removed, as unnecessary heating is harmful.

Similar points apply when plain perforated boards are used, and wiring is point to point underneath. Quite thin gauge wire can be used (say 28swg or so) for most circuits, with somewhat stouter wire (say 22swg) for those carrying current to small ICs, etc. Leads may often be arranged so that there are few cross-overs, and 1mm or other insulated sleeving can be put on connections where necessary. The wire ends of resistors and other components will often reach to the various connecting points.

Components to Use

Constructors sometimes seem to have difficulty in deciding on exactly what components to use. Notes on this should prove to be of help.

Resistors in all the circuits here can be ¼ watt or 1/3rd watt, 5 per-cent tolerance, except if a larger wattage is particularly listed as being required. In many places 1/8th watt or smaller resistors could be used while ½ watt or larger are in order, but take unnecessary space. All common values will be found in the E12 range and their decades – 10, 12, 15, 18, 22, 27, 33, 39, 47, 56, 68, 82; 100, 120, 150 . . . 1k, 1.2k, 1.5k . . . 10k, 12k, 15k etc.

Linear or log potentiometers may be indicated. With these, the ratio between rotations and resistance is different. One type will replace the other, but control effects may then be crowded over a small part of the adjustment. Generally, values such as 2k, 2.2k and 2.5k, 4.7k and 5k, 200k, 220k and 250k, or 470k and 500k (0.5 megohm) are interchangeable.

The voltage rating of capacitors should at least equal the actual voltage present. Thus if a 10v electrolytic capacitor is called for, a 12v, 15v, 16v or higher rating could be used, if to hand. Paper, mica, ceramic, polyester and similar small capacitors will usually have a voltage rating which is easily adequate, except occasionally in miniature types. Values such as 2µF and 2.2µF, 4.7µF and 5µF, 20µF and 22µF, 47µF and 50µF, are interchangeable.

For connecting purposes on circuit boards, tinned copper wire is convenient – about 20swg for power or return circuits, and 26swg for low level circuits. Some 1mm sleeving of two or three different colours will be handy for identifying leads. For external circuits, thin flexible wires are required. Use red and black for positive and negative respectively, with the correct battery clips. This will avoid damage from wrong polarity.

Screened single lightweight cable will be useful for audio leads, and 3.5mm jacks and sockets may be adopted throughout; or connectors will be chosen to match existing equipment. Flat twin low voltage flex is useful for speaker and other circuits.

CHAPTER 2

POWER SUPPLIES

Regulated Power Supplies

A number of ICs are ideal for use in regulated power supplies of the kind shown here. With these circuits, the output voltage is maintained at virtually the same figure, for large changes in the current drawn. This makes them excellent for all sorts of radio and electronic equipment, and the projects described here.

A PSU of this type will need only a few more components than does an unregulated supply. With the latter, the voltage fluctuates according to current taken.

Other regulators can be used than those shown, but must be of positive regulation type, for these circuits. One which is very useful, though more expensive than those listed, is the LM317K. This can be used for up to 30v 1.5A. It incorporates automatic overload and short-circuit protection, so that it or the PSU is not damaged by these conditions. It reverts to normal operation when the overload is removed. It is 3-terminal, with IN, ADJUST and OUT circuits.

When modifying circuits to make use of transformers or other items to hand, note that regulation and voltage control is only possible when the DC voltage supplied to the regulator device is somewhat higher than the output voltage required. When the voltage at the rectifier reservoir capacitor has fallen to near the PSU output voltage, as can arise with an inadequate transformer, regulation under full load cannot be maintained.

5v Supply for 7400 Series ICs

The series of 7400 TTL integrated circuits are intended to operate from a 5v supply, and integrated circuit regulators for this purpose are available. These maintain the correct DC output voltage, under a range of input voltages and loads, so are ideal for the operation of all sorts of 7400 ICs.

The supply shown here will provide power for digital ICs

needing up to 500mA. By constructing it as a separate unit, any circuit to be operated can be plugged in, and this allows it to provide power for various projects. It may also be used where a 5v 500mA supply is suitable, so can in some cases be pressed into service where a 4.5v or 6v battery would otherwise be fitted.

Figure 3 is the circuit. Transformer T1 provides isolation from the mains, and has a 9– 0–9v 500mA secondary. After rectification, the DC input voltage for IC1 is something over 9v.

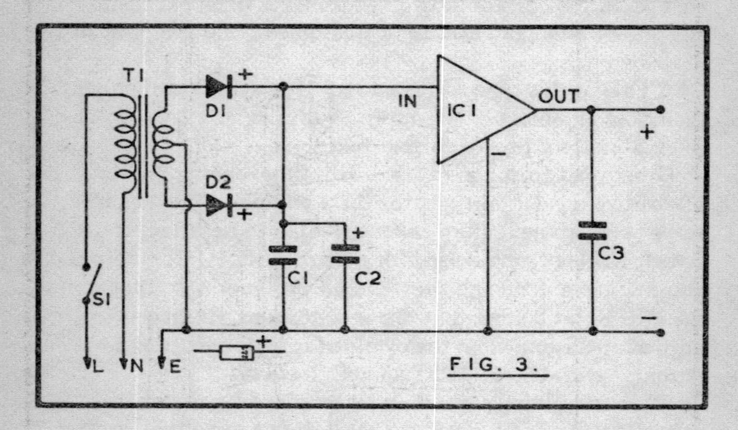

FIG. 3.

D1 and D2 are 50v 1A, these being readily available. C1 is to help suppress transients arising from the mains, and C2 is the main smoothing capacitor. Here, a value of 1500μF to 3500μF may be used. The higher capacitances provide better smoothing.

IC1 is the LM341P5. This is a 5v 500mA positive regulator. It has a TO202 case. Actual maximum current depends to some extent on other factors. Here, heat sinking is to the small metal chassis, and T1 is rated at 500mA, so maximum output is limited to 500mA for continuous load.

Numerous other 5v regulators are available. Some are of small current rating, to mount on individual circuit boards, while others are for heavy currents. When more current is required, T1 (and the rectifiers if necessary) should be of increased rating. Negative regulators are not suitable for the circuit.

Figure 4 shows layout on a small metal chassis. This can be a 4 x 6in "universal chassis" flanged member, as no components fit below. A tagstrip supports most of the components, and this is bolted by the brackets MC to provide metal chassis or negative circuit returns. Take 9v leads from T1 to negatives of the diodes, and diode positives to C2 positive.

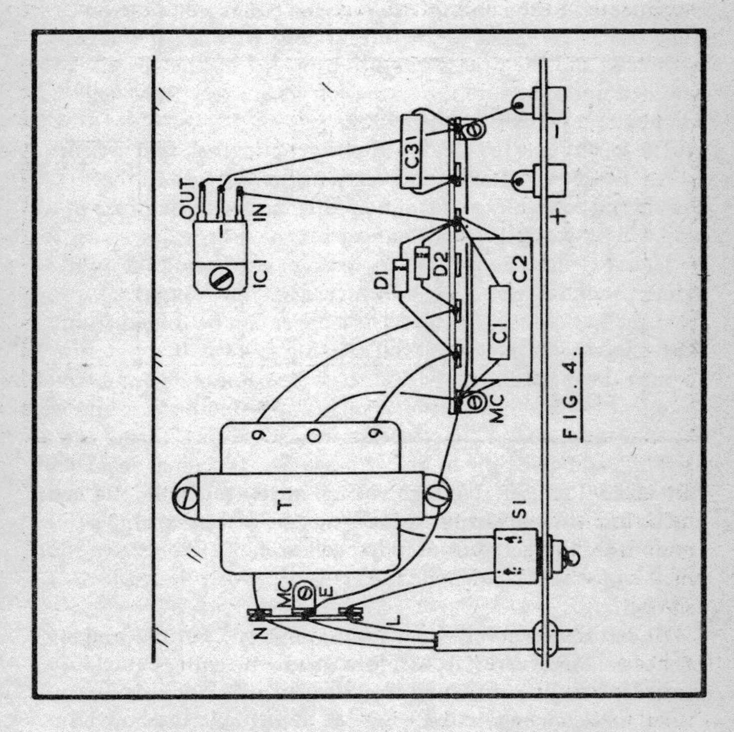

Drill a hole for a 4ba bolt for IC1, clear away burr so that it lies flat, and bolt it tightly in place. The legs can be bent up slightly, to clear the chassis.

Leads run from C3 to two sockets on the front of the power supply. Use red for positive.

Safety

The mains lead is 3-core cord, having yellow-green for earth, blue for neutral, and brown for live conductors. A thin cord (2A or 3A gauge) is adequate. Use a 3-pin fused plug, and place a low-rating fuse of 2A or 3A in this.

At the PSU, the cord passes through a rubber grommet, and its conductors are anchored on the 3-way tagstrip. Note that the switch S1 is placed in the live conductor L. The earth lead is connected to the chassis. This is also connected to the transformer core, and to the output negative line. If this lead is omitted and the negative return provided by the chassis is to be relied upon, be sure the fixing screws are very tight, and will not eventually become loose.

The earthing is for safety. If a short circuit should arise in T1 or elsewhere, which could carry mains voltages to the chassis and low voltage output circuit, the low-rating fuse in the plug will blow. This avoids danger for the user.

The PSU also needs to be enclosed in a case, so that no live mains circuits can be touched. A case may be bought, or made from further chassis members, or a cover can be folded from sheet metal, and fixed with self-tapping screws. It is not wise to omit the cover, once the PSU has been completed and tested.

Safety, with this and similar mains operated power supplies, depends upon sound construction, correct wiring, fusing and earthing, and shielding of high voltage points against accidental contact by the user. No high voltage must ever be able to arise on the low voltage output sockets, or low voltage wiring as equipment connected to the PSU will no longer be safe to touch.

Testing

A DC voltmeter should show approximately 5v at the sockets. Due to tolerances in IC1, and lack of exact accuracy of the meter, the reading might be very slightly above or below 5v.

To check for regulation, connect some item taking up to about 500mA. A 6.3v, 0.3A or 3w bulb is suitable. The reading at the output sockets should remain virtually unchanged.

Components:- 5v Supply for 7400 Series ICs (Figure 3)

Semiconductors

IC1	LM341P5		
D1	1N4001	D2	1N4001

Capacitors

C1	0.1μF 50v	C2	3500μF 16v
C3	0.1μF 50v		

Miscellaneous

T1 250v/9–0–9v 500mA Transformer
Mains toggle switch
Sockets, Case, Tagstrips etc.

1½–20v PSU

An adjustable, regulated power supply unit will prove to be of great utility. The PSU described here will deliver up to about 1 ampere, at any voltage likely to be required for receivers, calculators, amplifiers, and all sorts of other electronic items. The actual range is from about 1½v to 20v.

The integrated circuit regulator IC1 provides a stabilised output, the voltage remaining virtually unchanged, despite changes in the current drawn. Performance in this respect is of course very much better than that of the usual unregulated type of supply, where the voltage soars when the load is reduced, and drops when a heavy current is taken.

Figure 5 shows the circuit. Transformer T1 provides the necessary isolation from the mains, and low voltage, rectified by D1 and D2. A 20–0–20v 1A secondary is required, and this output voltage ought to be maintained at T1 secondary even when the full current is taken. After rectification, nearly 30v will appear across the reservoir capacitor C1.

Output is controlled by VR1, which allows setting of the "adjust" potential of IC1. The IC itself contains all the necessary semiconductor devices and circuitry for control of the output, in accordance with the setting of VR1. C3 helps remove ripple from the adjust input tag circuit.

M1 is a voltmeter, and indicates the output voltage. In use,

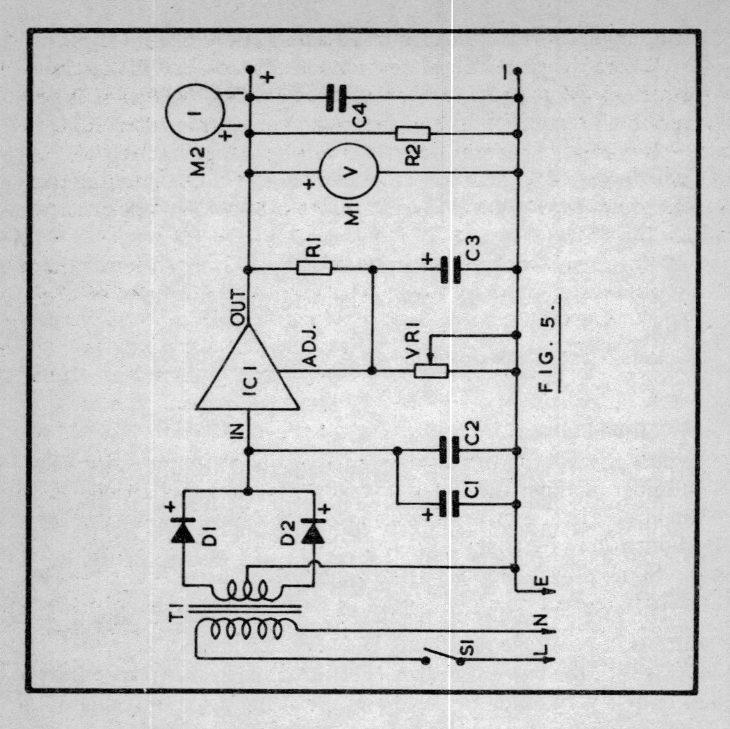

FIG. 5.

merely turn VC1 so that M1 shows the wanted potential. The equipment to be operated is then plugged in.

Small panel meters, with an internal resistor, suitable for M1, can be obtained. It is also easy to use a milliammeter here instead, by fitting a series resistor. The value of the resistor will depend on the meter. For a 1mA instrument, to read 0–20v, the value required is 20k. For a 2mA meter, use 10k. A 5mA meter will require 4k (2.2k and 1.8k resistors in series). Where convenient, a slightly different range can be provided, such as 0–24v, or 0–25v, to suit the meter scale. It would, as example, be awkward to read 0–20v on a meter calibrated 0–250, but if this were a 250μA instrument, 0–25v would be obtained by fitting a 100k series resistor, and so no new scale would have to be drawn. The value, for any meter and range, can be found from: Resistance = Voltage/Current.

Where the meter draws several mA, the resistor R2 can be omitted. Its purpose is to provide a small load when M1 is of very high resistance and no external apparatus is connected.

If wished, a current meter M2 can be placed in circuit. Here, a meter reading up to 1A can be fitted. A matching pair of meters for M1 and M2 will allow a professional appearance for the PSU.

More sensitive meters can be used for M2, by shunting them to obtain the required range. The shunt will normally be of very low resistance. Its value may be found from:-

$$\frac{Rm}{(N-1)}$$

where Rm is the internal resistance of the meter, and N is the number of times the full-scale reading is to be multiplied. As an alternative, a piece of resistance wire can be taken, and the shunt found by trial.

Some power supplies use a single meter, with 2-pole 2-way switch, so that it can be placed in series with a resistor, or across a shunt. It can then take the position of either M1 or M2, to read voltage or current, as required.

Construction

A case approximately 6 x 8 x 4in is convenient (152 x 203 x 102mm). Layout and wiring can be seen from Figure 6.

Remember to fit the mains lead and fuse in the manner described for the 5v PSU. This is to earth the case and negative of the low voltage line, for safety.

The power dissipated in IC1 generates heat, and this item is fitted to the back member of the case. The metal fixing plate is common to the output tag, so that means have to be taken to avoid a short circuit. This can be done by using a thin mica insulator under the IC, with a bush on the fixing bolt. The IC should lie flat, and the mica is thinly smeared with one of the thermal conductive pastes available for the purpose. An alternative is to fit IC1 to a plate or heat sink which is then secured to the back with bolts having insulated bushes and washers.

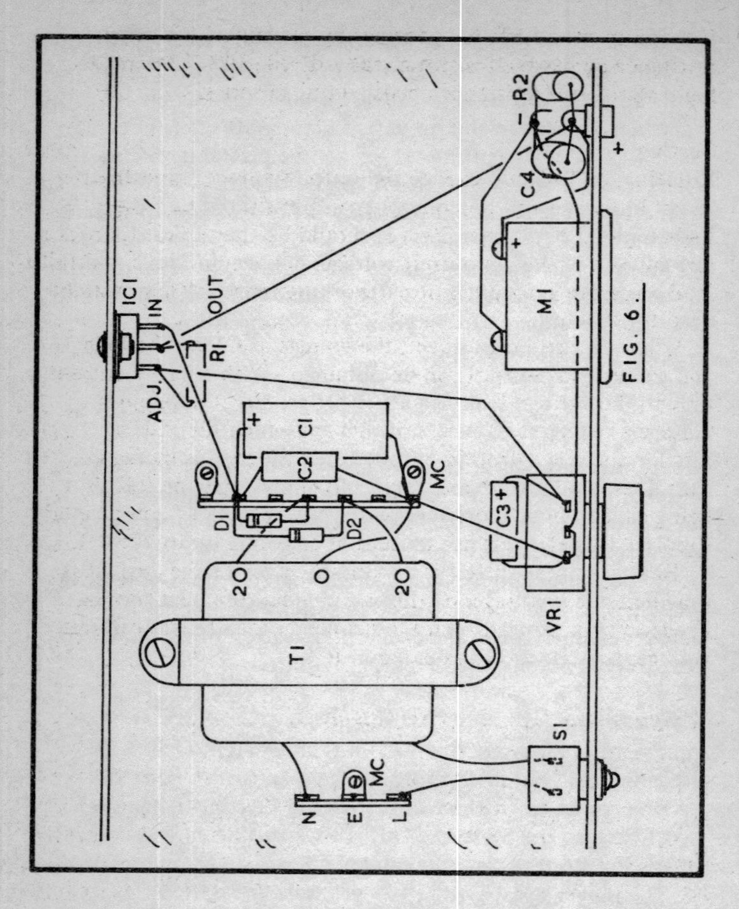

Connections inside the PSU should be of reasonably stout wire. At maximum current, a sensitive meter will show the voltage drop which arises due to leads and joints where the resistance is only a small part of an ohm.

If M1 is arranged to read voltage by using a series resistor, as explained, a check should be made with an accurate, external voltmeter. This will allow adjustment to be made, if necessary, to compensate for the meter resistance, or resistor tolerance.

If readings are a trifle low, correct by placing a high value resistance across the resistor already fitted. But if M1 reads high, additional resistance needs adding in series.

Testing

Rotating VR1 should change the output voltage smoothly. Exact limits depend on components, but should be from a little under 1.5v to over 20v. It should be found that there is virtually no change in output voltage, between no load and full load conditions, for the normal working range. A lamp can be used for this test.

When no current is drawn, the voltage at C1 will be high, and so over 20v output can be obtained. With a heavy current drawn, the voltage available at C1 drops. So at maximum voltage – which is in excess of that for which the PSU is rated – there is a drop in output when the load is increased. This does not indicate any fault, but shows that the PSU is being called upon to deliver greater power than T1 can properly supply. This effect is not present at anything up to 20v 1A.

Heat generated in IC1 is maximum when a large current is drawn at a low voltage. If this is too great, the heat sinking needs to be improved. With lower current, and higher output voltage, less power is dissipated in IC1.

Components:- 1½v–20v PSU (Figure 5)

Semiconductors

IC1	LM317T positive regulator		
D1	1N4001	D2	1N4001

Resistors

R1	220Ω 1W	R2	680Ω 1W
VR1	5k lin potentiometer		

Capacitors

C1	4700μF 30v	C2	0.1μF 50v
C3	10μF 30v	C4	0.1μF 50v

Miscellaneous

T1	250v/20–0–20v 1A transformer
M1	0–20v or 0–25v meter

M2 See text
Mains toggle switch,
Sockets, case, tagstrip etc.

2½A Adjustable PSU

The extra power obtainable from this larger power supply unit will be of utility to run car receivers, larger amplifiers and other equipment, as well as allowing the popular 10 watt 2 metre Amateur transceivers and similar items to be run from the mains. The PSU can provide up to 20v. It incorporates automatic current limiting, so that damage is avoided in the event of a short circuit of the output leads.

Figure 7 is the circuit. A neon indicator is included across the primary of the mains transformer T1. This consists of a small neon, in holder, with series resistor incorporated, and is a useful addition to show that power is on.

T1 has a 20v 3A secondary, and four rectifiers are fitted in a bridge circuit. A centre-tapped secondary is not required with this arrangement. A 50v 3A bridge rectifier can be used instead of the individual rectifiers. C1 is the reservoir capacitor, with C2 to suppress transients.

IC1 is the regulator and means of voltage control. Input to IC1 is at 2 and 3, and output is taken from 8 to a large current passing transistor Tr1. C3 and C4 are suppressor capacitors.

The output at 8, and consequently PSU output voltage, depends on the setting of VR1. R2 and R3 are included so that point 6 can never be taken to full negative or positive lines. When wiring and setting VR1, note that voltage *falls* as VR1 wiper is moved towards R2, as this may be the opposite to that expected. This method of control helps stabilise output, as a slight rise in output voltage or vice versa operates to provide automatic compensation.

Current limiting, or the output current at which IC1 closes down to protect the transformer and rectifiers, depends on the resistor R1. With a 10 ohm resistor here, it was found that current limiting was at 3 amperes. However, spreads in IC1, and slight changes in internal resistance in wiring, may modify this. If the maximum current is too low, R1 should be

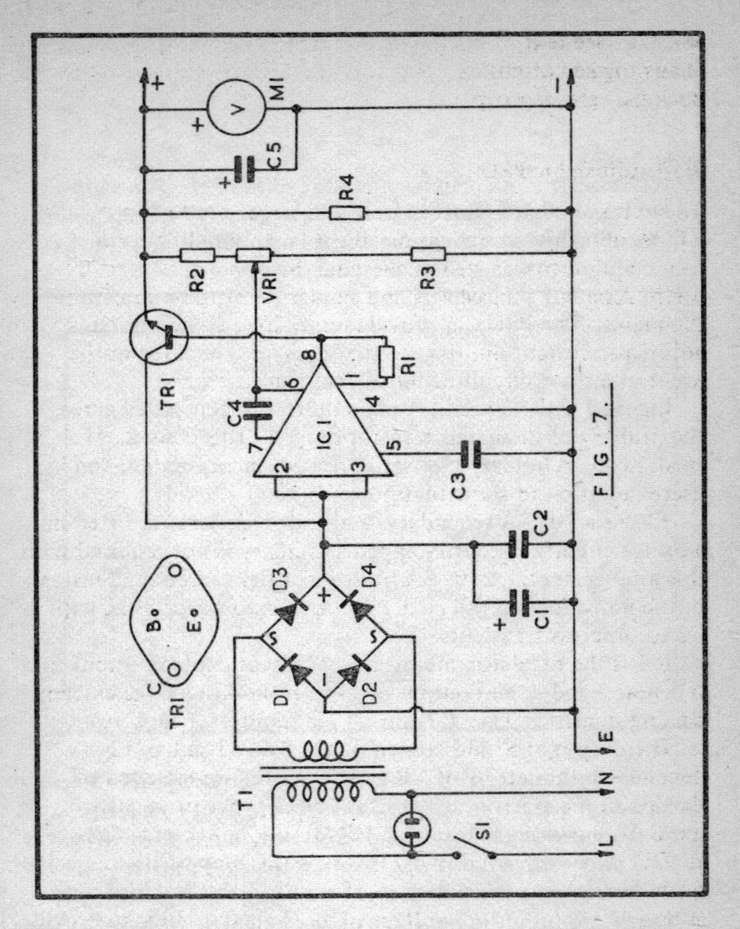

FIG. 7.

reduced in value – say to 8.2 ohm. Alternatively, if current is too heavy with the output shorted, R1 can be increased slightly in value, until this is corrected.

R4 provides a small load, with no external equipment connected. M1 is a voltmeter, 0–25v or 0–30v. If M1 draws appreciable current, R4 can be omitted.

As the integrated circuit IC1 cannot itself handle this

current, the IC controls the base of Tr1, which in turn controls output, as the emitter voltage follows the base voltage, derived from the regulator.

Construction

General construction can be along the lines of the PSUs already described, though a larger case will be necessary to accommodate T1. General layout can follow Figure 6 and Figure 7B shows

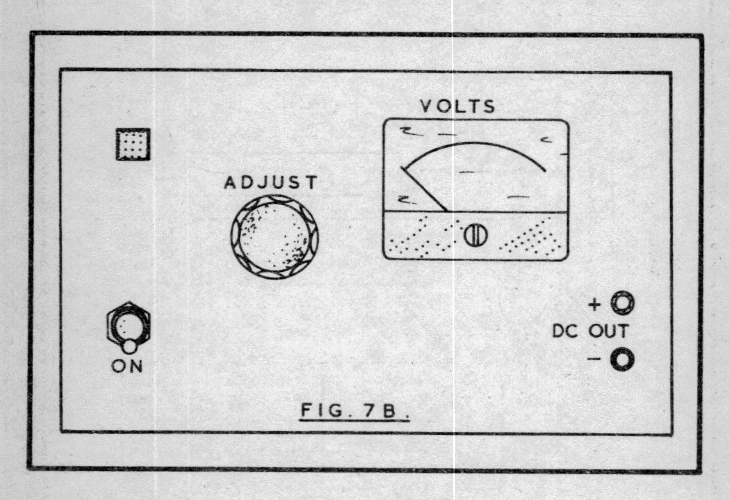

FIG. 7B.

the panel with meter, VR1, etc. Figure 8 shows the rectifiers mounted on a tagstrip, which also supports C1. Two leads come from the secondary of T1, as shown, and a positive lead goes to the positive line of the control board. Positive of C1 is also connected to Tr1 collector.

The board is 0.15in matrix, and can be wired, or have vertical foils. These form the connection shown. Elsewhere, the foils must be cut to avoid short circuits. That is, under C2 and R2, and under IC1. IC1 has eight leads, which must of course be correctly positioned. A small clip-on heat sink is fitted to IC1.

Flying leads are provided for positive, base and VR1. The negative line is made by means of the tags at the mounting

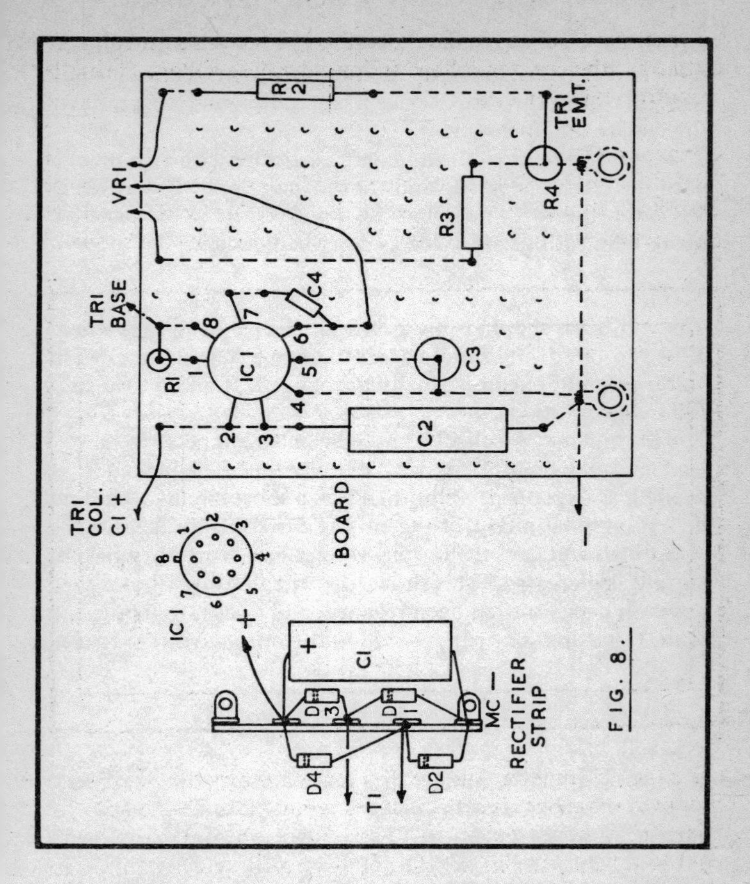

FIG. 8.

holes, as well as by a lead. Those leads which carry a high current, in particular, must be of stout wire. They are from T1 secondary, rectifiers and to C1, from C1 to Tr1 and output, and from C1 negative to output.

Tr1 requires a heat sink. This can be the metal case, or a separate finned sink about 3 x 5in (76 x 127mm) mounted with its fins vertical. Tr1 is insulated from the case (or sink) by means of an insulation set, consisting of shaped mica washer, and bushes for the bolts. Smear heat conducting grease on the washer first, and avoid any burr or unevenness which will prevent Tr1 lying flat. (An alternative is to insulate the whole heat sink from the case). Place a tag under one fixing nut, for the collector connection.

It is convenient to fit the board vertically with two brackets, with the rectifier tagstrip and transformer near. The meter and VR1, together with neon indicator and S1, fit to the panel. Mains connections are made as described earlier.

Testing

The voltmeter should show a smooth change in voltage when VR1 is rotated, and should agree with an external meter. The voltage should remain virtually unchanged, from no load to full load conditions.

Limiting may be checked by placing an ammeter in one lead, and shorting the output. R1 may then be adjusted, if needed, as explained. In normal use, a short circuit would not be left on indefinitely, as the cut-off current is quite large.

At high voltages, the output voltage can drop when output current is increased. This shows that the limit of T1 and the reservoir capacitor has been reached, and does not denote a fault. This does not happen within the normal voltage range.

Components:- 2½A Adjustable PSU (Figure 7)

Semiconductors

IC1	CA3085 with heat sink		
Tr1	2N3055 with heat sink		
D1-D4	4 x 50v 3A rectifiers.		

Resistors

R1	10Ω ½W	R2	8.2k ½W
R3	1k ½W	R4	680Ω 1W
VR1	2.5k lin potentiometer		

Capacitors

C1	3500μF 30v	C2	0.22μF 50v
C3	0.1μF 50v	C4	47pF or 50pF
C5	10μF 30v		

Miscellaneous

T1	250v/20v 3A transformer
M1	0–25v or 0–30v meter

Neon indicator, mains toggle switch,
Sockets, case, tagstrips, board, etc.

CHAPTER 3

RADIO AND AUDIO PROJECTS

ZN414 Receivers

The ZN414 integrated circuit will be found extremely useful for the construction of tuners and receivers of many kinds. Although this IC is contained in a TO18 package, of the same size as many small transistors, and has only three leads, it contains ten transistors. These provide radio frequency amplification, detection and automatic gain control. Input to the IC is from a tuned circuit, which will select the wanted transmission. Output from the IC is audio frequency, at sufficient power to operate phones, or ready to pass to an audio amplifier.

In view of the great simplification made possible by using this IC, is is not surprising that it finds such ready use in receivers. Typically, it operates from 1.3v, the actual supply voltage range being 1.2v to 1.6v, and current is about 0.3mA to 0.5mA.

It will be noted when using the ZN414 that a small change in the operating voltage can produce quite a substantial modification to results. When the voltage is rather low, sensitivity falls off, and a smaller audio output is obtained. With an increase in voltage, sensitivity rises so that more transmissions will be tuned in, and volume rises considerably. However, instability in the form of whistles or oscillation may arise at some frequencies, so that more than 1.5v is often impracticable. For midget receivers, it is satisfactory to use a small mercury battery, or single ordinary dry cell (1.4v). With larger receivers, a suitable low voltage is easily obtained by means of potential dividers, or Zener diode.

Phone Portable

Figure 9 is the basic receiver circuit. L1 is the ferrite rod aerial, tuned by variable capacitor VC1. C1 and C2 are by-pass capacitors. R1 is 100k and provides feedback from output

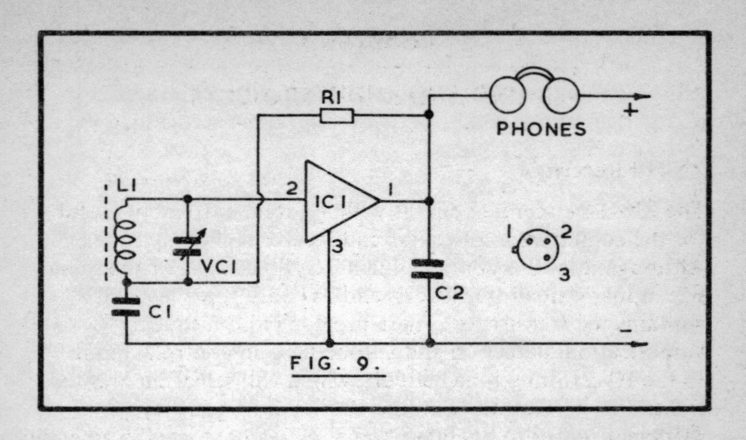

FIG. 9.

to input, for automatic gain control purposes. Leads are input (2), output (1) and earth (3) as shown.

L1 can tune the usual medium wave band, or around 200–600m or 1500–500kHz. The medium wave winding from a disused or broken receiver can be utilised, with a capacitor of about 208pF. There is considerable latitude, as no ganging with other tuned circuits is necessary. If L1 is to be wound, then 90 turns of 26swg enamelled wire can be used on a thin card tube, arranged to slide on a 3/8in (9½mm) diameter rod. Longer rods slightly increase signal pick-up, so 6in or 9in (152–130mm) rods can be used with a non-miniature receiver. For a miniature receiver, a much shorter rod will be sufficient. VC1 can now be 300pF or 365pF. Coverage can be altered by moving L1 on the rod.

C1 can be 10nF, and C2 is 0.1μF. When a headset of about 500 ohm resistance is to hand, very good results can be expected. The phone windings complete the positive circuit, and provide the output circuit load for IC1. If they are connected by means of a miniature jack plug, the receiver is automatically switched off when this is pulled out of its socket.

In other circuits, or when using a crystal earpiece, the load is provided by a resistor. This is generally 470 ohm, though a higher value may be used in some circuits. Audio signals are then taken from 1 of IC1 by means of a coupling capacitor.

The low voltage supply is convenient for the IC alone, or with a simple transistor amplifier. When a larger amplifier or IC audio amplifier will be added for speaker reception, a 9v battery is more suitable.

Construction

The circuit can be assembled in a plastic box, as in Figure 10. This can use a 5in (127mm) or similar ferrite rod, and no attempt is made to produce an extremely small or compact receiver.

With this and other ZN414 circuits, C2 needs to be close to lead 1 and 3. Lead 2 should be reasonably clear from lead 1. The device has exceedingly high gain, and any unwanted feedback will cause the instability described.

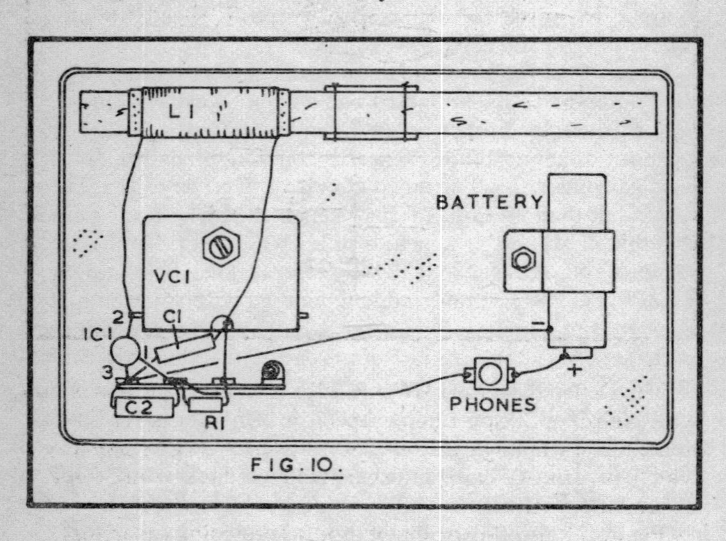

FIG.10.

VC1 can be a miniature air-spaced component, with advantage, where space allows. Details of L1 have been given. Any ordinary single cell can be used, and leads may be soldered directly to this. The end cap is positive. No on-off switch is fitted, and the headphones are plugged in to complete the battery circuit, as explained. Cut a small block of wood to support the ferrite rod, and secure this with cotton or elastic.

Volume will normally be adequate, unless the receiver is used in a poor or screened locality.

In some receivers it may be wished to add long waves. This can be done by using a LW winding on the rod, Figure 11. L1 is the usual MW winding, and L2 for long waves. When S1 is closed, the latter is shorted out. Turns of L1 and L2 should be in the same direction. If this is not very clear from examining the coils, the correct way for L2 can be found by reversing it, or its connections, if necessary.

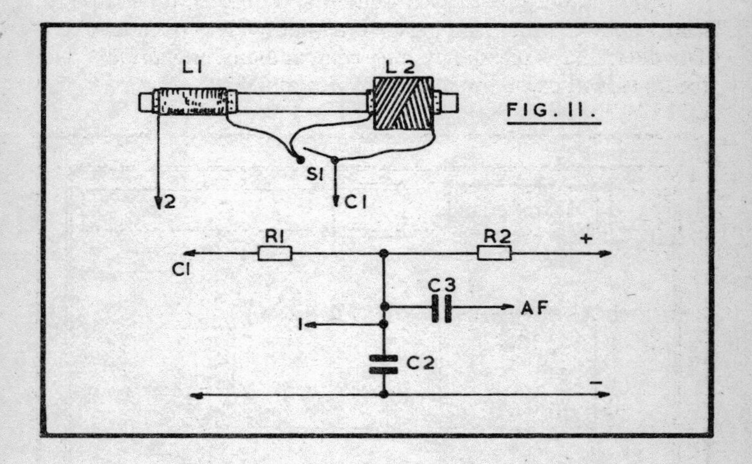

The Denco (Clacton) MW/LW FR5 is suitable for this circuit, with 200pF or 208pF tuning capacitor. An alternative is a home-made winding of about 250 turns of 38swg or similar wire, between two cards or other insulated discs, with 300pF or similar capacitor.

Figure 11 also shows the method of providing capacitor coupling. IC1, with R1 and C2 are wired as originally, but R2 replaces the phones, and is 470 ohm. C3 can be 0.47μF to 0.1μF, and couples audio signals to the following circuit. If a screened lead is provided, with inner conductor to C3 and outer braid to negative line, signals may be taken to an audio amplifier, and the receiver will then operate as a tuner. An on-off switch should be added in one battery lead.

A crystal type earphone can also be fed from C3. However, for general listening a medium or high impedance headset, with headband, will be more convenient and satisfactory.

Components:- Phone Portable (Figures 9 & 10)

Semiconductors

IC1	ZN414

Resistors

R1	100k

Capacitors

C1	10nF	C2	0.1μF
VC1	208pF or 300pF or 365pF see text		

Miscellaneous

L1	see text
Phones	500Ω impedance
Battery	see text

Additional Components for LW (Figure 11)

L1/L2	see text
R2	470Ω
C3	0.47μF to 0.1μF

Miniature Receiver

Figure 12 shows a very small receiver which uses the same circuit. VC1 is a 120pF compression type capacitor, and its adjusting screw is removed, and replaced by a longer screw to which a small knob is locked. The knob is 10mm in diameter. Replace the washer after putting a small nut between the knob and capacitor, so that the capacitor can be opened and closed despite the thickness of the case.

Construction is on a piece of perforated board about 32 x 22mm. Solder the tags of VC1 to two pins inserted in

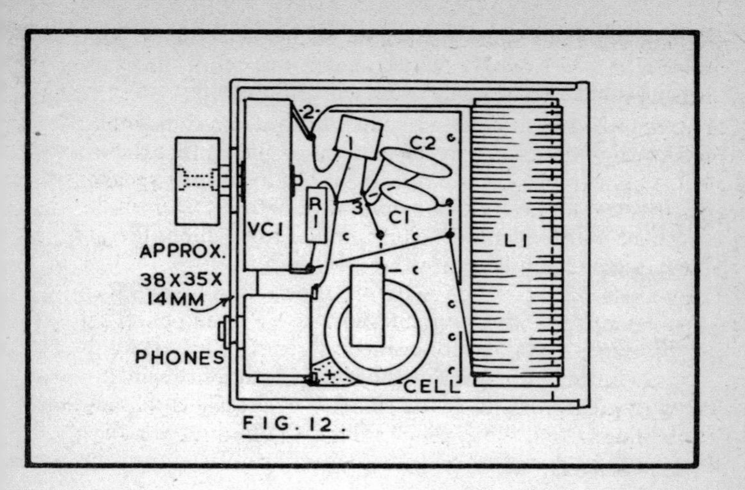

FIG. 12.

the holes, and snip off the projecting pins below the board.

The bottom of the mercury cell is positive. This rests on a piece of thin metal soldered to one jack socket tag. A strip is bent to press on the top of the cell, and passes through a slot in the board, where it is fixed by soldering it to a wire coming up through two holes.

IC1 and the other components are fitted in the space behind the variable capacitor, located about as shown. The ferrite aerial rod is 33 x 9½mm. A piece can be cut from a longer rod by filing a notch all round, then snapping off with the fingers. L1 is 95 turns of 38swg wire. Full medium wave coverage cannot be obtained with VC1, so it is necessary to curtail coverage a little. The number of turns on L1 can be changed, if wished, so that the coverage provides reception of the more important frequencies.

The case is part of a mint sweet box, cut down in length. A hole is drilled to clear the nut operating VC1, and for the socket. The drills must be sharp, and used with only light pressure. Take off the jack socket nut, and remove VC1 operating knob and screw, holding the board vertical so that the washer of VC1 is not displaced. Put in the case, and insert the knob and screw, and fix the socket. A little is also cut off the lid, which fits inside the case. Other small boxes could be used. It is not easy to reduce dimensions below those given, using standard and easily obtained components.

Components:- Miniature Receiver (Figure 12)

Semiconductors
IC1 ZN414

Resistors
R1 100k

Capacitors
C1 10nF small disc ceramic
C2 0.1μF small disc ceramic
VC1 120pF compression type

Miscellaneous
L1 see text
2.5mm miniature jack socket, board, mercury cell, case, etc.

2 IC Receiver

To obtain loudspeaker reception, an audio integrated circuit offering a fairly high degree of amplification can follow the ZN414. This combination makes a good portable receiver for home use.

In Figure 13 IC1 is the ZN414, and L1 covers medium waves, and is tuned by VC1. Long wave coverage can be added in the way described. R1 is the feedback resistor for gain control, and R2 the load, with C2 for by-pass from 1 of IC1.

The supply for R2 is derived from the junction of VR1 and R3, which form a potential divider. If wished, VR1 may be replaced by a fixed resistor of 470 ohm, 5 per-cent, and R3 by a resistor of 2.7k. With the receiver running from a 9v battery, the supply point will then by approximately correct. However, there is some advantage in fitting a small pre-set potentiometer as shown, as adjustments can then be made to compensate for component tolerances, for optimum performance. C4 is a decoupling or by-pass capacitor.

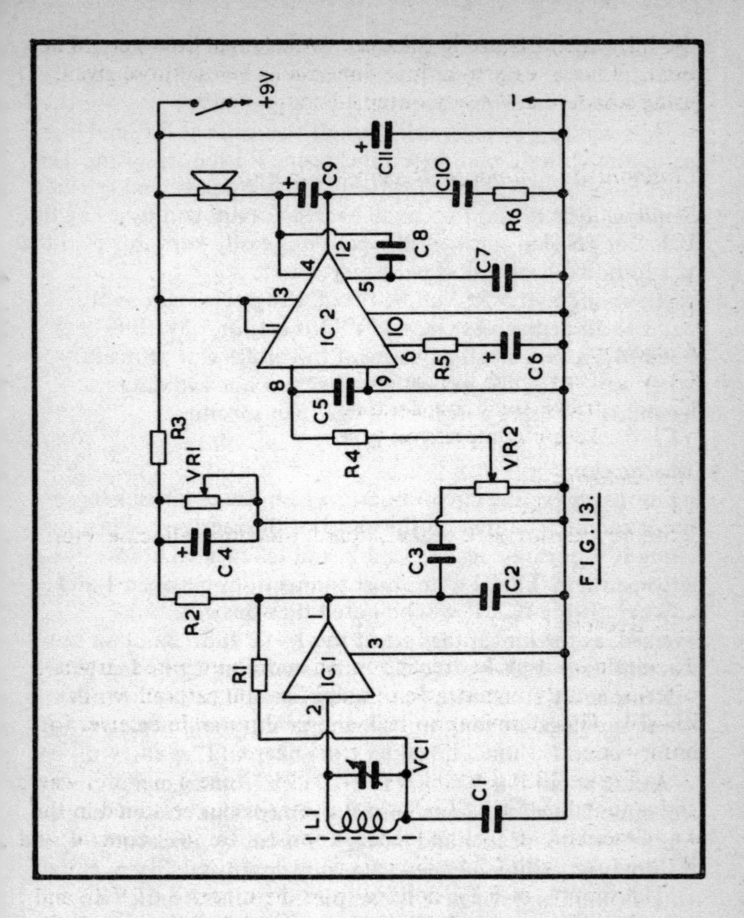

FIG. 13.

Audio is taken from 1 of IC1 by coupling capacitor C3, to the volume control VR2. This is panel mounted, and incorporates the usual on-off switch. R4 is a parasitic stopper.

IC2 is the TBA800. With a 9v supply, this IC can deliver up to 1w output with a 6 ohm speaker, or about 800mW with 8 ohm and 500mW with 16 ohm units. Under 6 ohm should not be connected. The output obtained with 8 ohm or 16 ohm speaker is easily adequate, and there is some saving in battery power. Current drawn depends directly on the volume, and is typically under 10mA with no signal or low volume, rising to 100mA or more for high output.

Gain can be preset by selection of the value at R5, and is maximum with 47 ohm here. Increasing R5 reduces gain. The other components are for load equalising, by-pass, and speaker coupling. The IC has two small heat sink tabs, and by using the 8 ohm or 16 ohm speaker, dissipation is easily kept down so that no additional heat sink is necessary.

Assuming that VR1 will be fitted, a high resistance voltmeter is clipped across C4, and VR1 is set for 1.3v. Subsequently, a very small adjustment to voltage can be made either way. This allows best sensitivity, while avoiding instability, or using a higher voltage than specified.

Construction

Figure 14 shows the circuit board, which has two brackets so that it can be mounted on the panel of the receiver. When fitting IC1, separate leads 1 and 2, and take lead 3 across between them. Fit C2 with short connections between 1 and 3.

When fitting IC2, it will be noted that this cannot be reversed, as the longer tags are at the 1–12 end. Bend up the heat sink tabs slightly, to clear other items on the board.

Reasonably stout wire (say 20swg) should be used for the power and negative lines to IC2, and positive and negative battery circuits should be taken from near C11, as shown.

A case about 6 x 5 x 3in (152 x 127 x 76mm) can accommodate all the items, but the dimensions chosen may depend on the speaker and battery chosen. A miniature 9v battery is not recommended. For maximum stability, a metal panel is used, and negative is completed to this by the two brackets. However, metal is not essential. If the case is wholly insulated, take a lead from the fixing bush of VR2 to the negative line connection near C5.

With the metal panel, VC1 must be insulated. With the air spaced type of capacitor which is secured with three 4ba bolts, this can be arranged by drilling the holes larger than required, and using insulated washers between capacitor and panel, and under the screw heads. The spindle passes through a clearance hole.

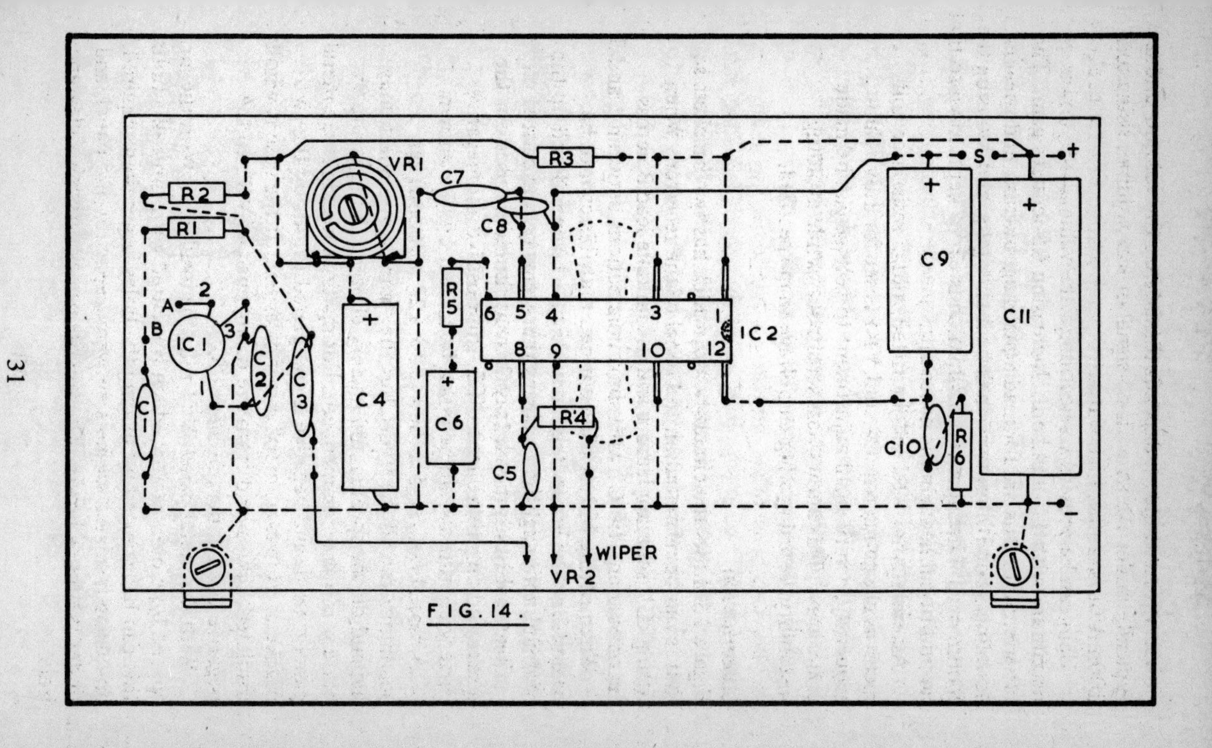

R2
R1
VR1
R3
C7
C8
A
2
B
3
IC1
1
C2
C3
C4
R5
C6
6 5 4 3 1
8 9 10 12
IC2
R4
C5
C9
C11
C10
R6
S
+
-
WIPER
VR2

FIG. 14.

The ferrite rod is mounted towards the top and back of the case, with a wooden block and thread or elastic. Leads from it run to VC1. Connect the frame (moving plates) of VC1 to B, and the fixed plates tag to A on the board.

Run the positive lead to the switch on VR2, and to a positive connector. Take a twin twisted lead from points S on the board to the speaker, keeping these wires away from the ferrite aerial winding and VR2. It is best to have battery and speaker under the board. With a reasonable placement of wiring, chances of instability are unlikely.

No adjustments are required, except for setting VR1 in the way described. If necessary, band coverage can be modified by altering the position of the winding on the ferrite rod.

It is worth noting that the TBA800 can be operated with up to 24v. This means that the receiver can be used with an existing mains power supply of more than 9v, if R3 is increased in value. As the small speaker will not be able to handle the audio power which would become available, a supply of up to about 12v to 15v maximum is most suitable. Alternatively, the receiver can be run from an adjustable power supply, such as described elsewhere. If this is set to provide 9v, it can be interchanged with the battery without adjustments.

The ferrite rod aerial has some directivity, in the usual way, and occasionally it may be necessary to turn the receiver, for best results from the wanted transmitter.

Components:- 2 IC Receiver (Figure 13)

Semiconductors

IC1	ZN414	IC2	TBA800

Resistors

R1	100k	R2	1k
R3	2.2k	R4	2.2k
R5	56Ω	R6	1Ω
VR1	500Ω pre-set	VR2	100k log potentiometer with switch

Capacitors

C1	10nF	C2	0.1μF

C3	0.1µF	C4	6µF 25v
C5	100pF	C6	100µF 25v
C7	2.2nF	C8	330pF
C9	320µF 25v	C10	0.1µF
C11	1000µF 25v		
VC1	300pF variable		

Miscellaneous

L1 MW ferrite rod aerial

3½ (90mm) 16Ω speaker, board, case, battery, etc.

Solar Radio

A receiver which will operate when powered by daylight, or by reasonably strong artificial light, can be made with the circuit in Figure 15. This uses the ZN414 integrated circuit, which has been described, and audio is available at the output lead 1. As strong signals tend to overload the audio amplifier Tr1, VR1 is fitted as an audio gain control.

VR1 could be 500 ohm, and replace R2. However, this value does not seem readily available in the type of miniature

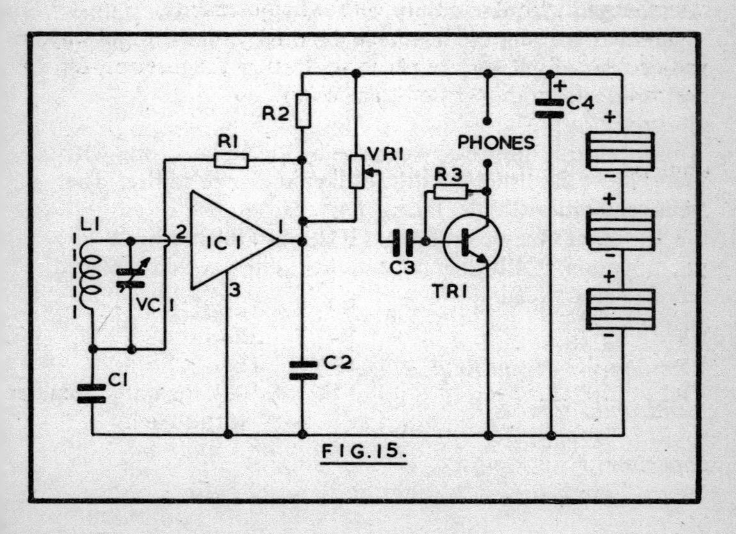

FIG.15.

manual potentiometer which is most convenient, so a higher value is used, and has to be shunted by the fixed resistor R2. If VR1 is omitted, take C3 directly to the output point of IC1. Some protection against over-loading can be obtained by turning the receiver to reduce the signal pick-up from local transmissions.

Output from Tr1 is for medium or high impedance phones. A pair with headband will be most comfortable. A good headset will provide better volume than a miniature earpiece.

Power is obtained at about 1½v from three silicon photo-voltaic cells in series. The Ferranti FRB150 cells were used here, but no doubt other cells of equal efficiency would be suitable. C4 is necessary to remove the hum, from modulation by interior lighting, which is otherwise experienced. This is so even with ordinary domestic filament lamps, where the effect may not be seen. With daylight, no such effect arises.

Volume depends somewhat on the degree of lighting. Indoors, and particularly with artificial light, the receiver can be placed in such a position that a steady, suitable voltage is produced. Outside, effects of this kind depend somewhat on the weather. They could possible be reduced by arranging for a more than adequate supply, and reducing this to a stable figure with a Zener diode, but this is omitted here in the interests of simplicity.

Construction

This receiver is most easily constructed in a plastic box with transparent lid, using the type of layout shown earlier. The cells can then be inside, facing upwards.

The circuit may be run from a 1½v dry cell, without modification. C4 is then not required, but a switch is placed in one battery lead.

Components:- Solar Radio (Figure 15)

Semiconductors

IC1	ZN414
Tr1	BC108, 2N3706, etc.

Resistors

R1	100k	R2	560Ω
R3	1.8M		
VR1	5k edge control potentiometer		

Capacitors

C1	10nF	C2	0.1μF
C3	0.1μF	C4	60μF 6.4v
VC1	variable to suit L1		

Miscellaneous

L1 MW ferrite rod aerial
Jacket socket, board, tagstrip, case, etc.
Solar cells, see text.

IC Tuner-Preamplifier

This tuner can be used alone to operate headphones, and can then run from a battery. Or it may be used in conjunction with the audio amplifier and power supply, shown later, or may be arranged to operate from whatever voltage may be available.

L1, Figure 16, can have seventy turns of 32swg wire, on a ferrite rod 3/8in (9½mm) in diameter, and about 4in to 5in (102–127mm) long. It is of advantage to wind the turns on a card sleeve, so that it can be moved on the rod to modify the band covered. The ends of L1 may be held with touches of adhesive, but the whole winding must not be waxed, painted, or otherwise treated. With this winding, VC1 can be 300pF to 365pF.

A higher Q will be obtained with a commercially manufactured coil which is wound with Litz wire (this has individually insulated strands reducing the RF resistance). A suitable coil can be purchased, or may be found in a broken receiver. VC1 is often 208pF for these aerials, though this exact value is not essential here.

Operating voltage for IC1 is obtained through R2, from the wiper of VR1. This allows setting for any line voltage which may be present. If wished, a fixed resistor may be included between the top of VR1, and positive of C5, as only about 1.5v will be required. Initially set VR1 wiper near the negative end,

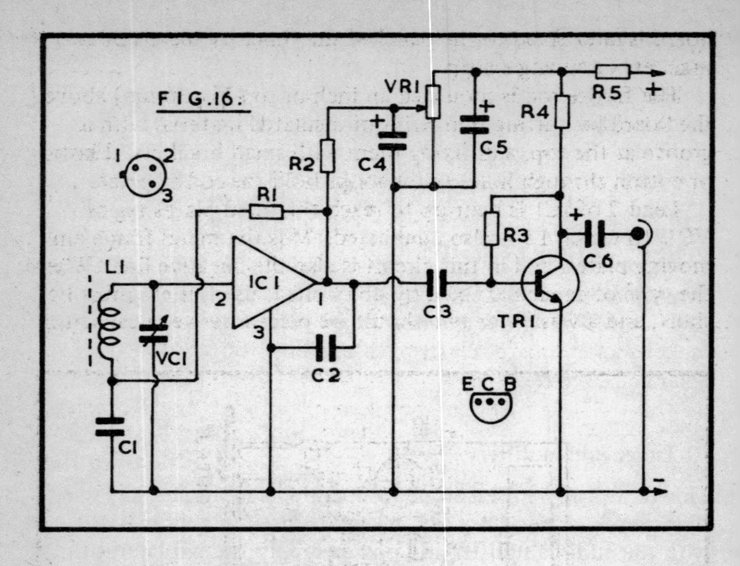

so that voltage to IC1 is near minimum.

Audio is available at 1 of IC1, and C3 couples this to the preamplifier Tr1. Here, base current is from R3, and R4 is the collector load. C6 is the output coupling capacitor, to isolate audio connections.

High or medium impedance headphones may be plugged directly into this outlet. Or a screened lead can be used to take audio to the main amplifier, which should have its own volume control. If wished, an audio gain control could be included in the circuit from C6.

If battery running is required, R5 and C5 may be omitted, and a 9v battery substituted. An on-off switch is placed in one lead. No switch is included in the tuner, as it is intended for operation from the amplifier described.

Note that R5 may be changed if necessary, to obtain about 9v or so across C5. The tuner can then be used with an amplifier having a 24v or other higher voltage supply.

Construction

A layout on perforated board is shown in Figure 17. This

board is later fixed to the panel of the tuner by means of two brackets or an angle strip.

The ferrite rod is mounted an inch or so (25–30mm) above the board by cutting two strips of insulated material with a groove at the top, and fixing them with small brackets. Elastic or cotton through holes in the strips hold the rod in place.

Lead 2 of IC1 is bent up to reach the fixed plates tag of VC1, to which L1 is also connected. M is the metal frame and moving plates, and in this circuit is also the negative line. Where the type of capacitor fixed by 4ba bolts is used, these must be short, and a washer or two should be placed between capacitor

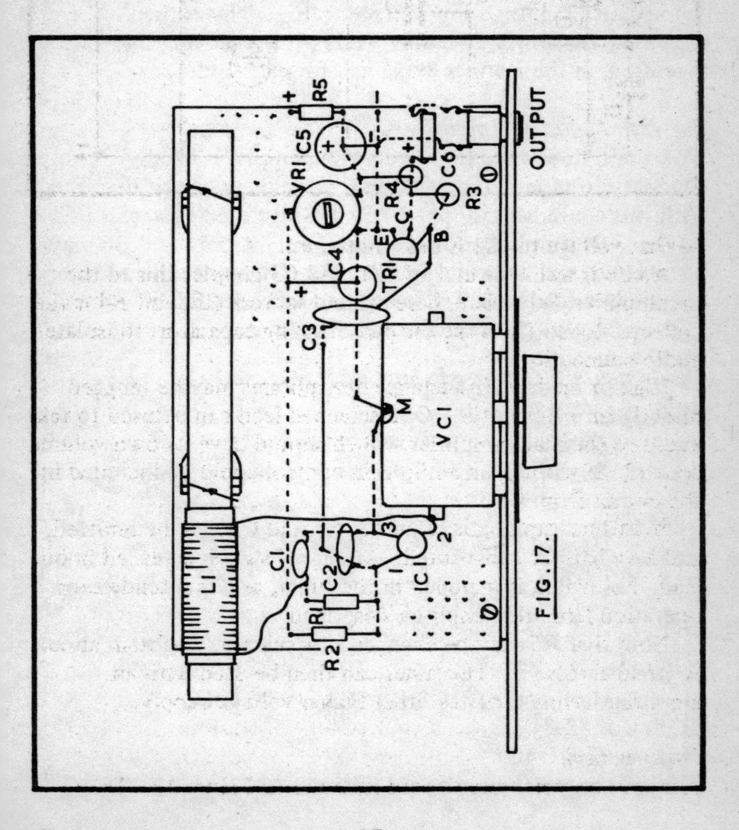

FIG. 17.

and panel, or the capacitor may be shorted. Fit an adhesive scale, or knob with dial, for tuning.

The exact layout is not too important. Some of the resistors and the larger capacitors are fitted vertically. When connecting the output jack socket, take the outer or sleeve contact to the negative line, and tip contact to C6. A 3.5mm outlet is convenient.

Solder red and black flexible leads to positive and negative (R5 and earth line) and fit a non-reversible 2-pin plug to suit the power supply and main amplifier power outlet.

Remember that VR1 is initially set with its wiper quite near the negative tag, for a low voltage to IC1. This is then increased, by slowly adjusting VR1, while checking the voltage across C4, in the manner explained for earlier circuits.

Use with Mains IC Amplifier

With the LM380 amplifier and power supply, it is only necessary to plug in the audio lead and power circuit described. With this equipment, about 11.5v can be expected across C5.

If a little simplification is sought, Tr1, R3, R4 and C6 may be omitted, and audio taken from C3. Otherwise the additional volume obtained with Tr1 will generally be useful.

Components:- IC Tuner-Preamplifier (Figure 16)

Semiconductors

IC1	ZN414		
Tr1	2N3706		

Resistors

R1	100k	R2	560Ω
R3	100k	R4	4.7k
R5	1k		
VR1	2.5k pre-set		

Capacitors

C1	10nF	C2	0.1μF
C3	0.47μF	C4	10μF 6v
C5	470μF 12v	C6	1μF 10v
VC1	see text		

Miscellaneous
L1 see text
Board, case, etc.

Mains IC Amplifier

A complete audio amplifier, with internal power supply for AC mains, is extremely useful. The LM380 integrated circuit audio amplifier needs very few external components, and a circuit for it is shown in Figure 18.

T1 isolates low voltage circuits from the mains, and has a 12–0–12v secondary. After rectification by D1 and D2, about 15v to 18v will be found across the reservoir capacitor C3, depending on the load. IC1 has internal protection against short-circuit of the loudspeaker or output, and also temperature protection. This means that damage is not caused by inadequate heat sinking, or excess drive. Operation returns to normal when the cause is removed. An output of over 2 watts can be obtained, with a heat sink. With no heat sink except that provided by the circuit board, this output cannot be maintained, but power handling is easily adequate for ordinary purposes in the home or for similar use.

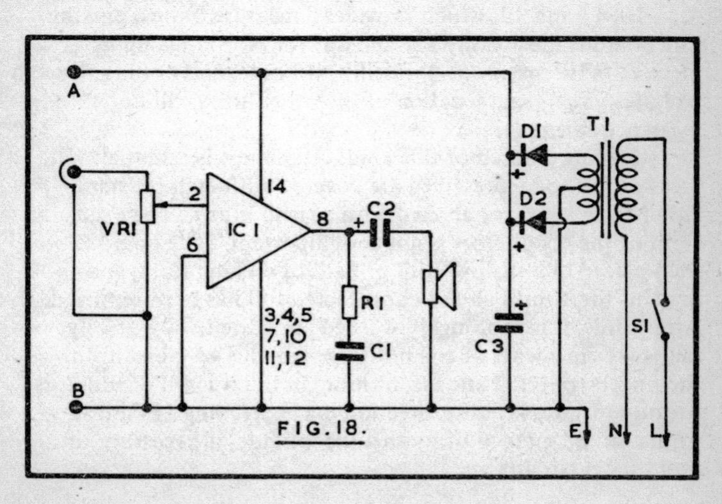

FIG.18.

Battery operation of this amplifier is feasible. For this, omit T1, D1 and D2. The battery can be 9v to 18v. A 9v or 12v supply will be convenient, either from a non-miniature 9v battery, or from cells in holders. Six 1½v cells will be needed for 9v, and eight cells for 12v. Actual current drain depends on the voltage, and volume. It is under 20mA with no signal or modest output, rising to 100mA or more for 500mW output peaks.

Sockets A and B allow positive and negative (earth) leads to be taken on to other units, such as a preamplifier or tuner.

Audio signals are taken by a screened lead to the audio socket and volume control VR1. The gain of this IC is not so high as that obtainable with some audio ICs having a preamplifier, so it is not suitable for low level inputs. With these, a separate preamplifier should be used.

The speaker may be 4, 8 or 16 ohms. Under 4 ohm should not be used. For this circuit, the 8 ohm impedance is most suitable. But somewhat similar speakers can be satisfactory.

Construction

Figure 19 shows the underside of a printed circuit board for the amplifier and power supply. This has an area of foil for 3, 4, 5, 10, 11 and 12, which provides a measure of heat sinking. The board is most easily marked out for IC1 by clamping a piece of 0.1in matrix perforated board here, and drilling through its holes. The exact location of the other holes will not influence assembly.

In making a board of this kind, cleanliness is essential. The areas of foil to be preserved are covered with etch-resistant – either from a pen, or applied with a small brush. The exact form of the conductors is not too important. The board is then placed in a shallow dish or bath of etching fluid, and remains there until all the unprotected foil has been removed. Afterwards, it is thoroughly washed, and dried. Most likely faults are fragments of foil not removed, due to grease from the fingers, or resist streaks; or holes in the wanted conductors because of insufficient protection during etching. Foil fragments can be scraped off with a sharp blade, if necessary to avoid short circuits.

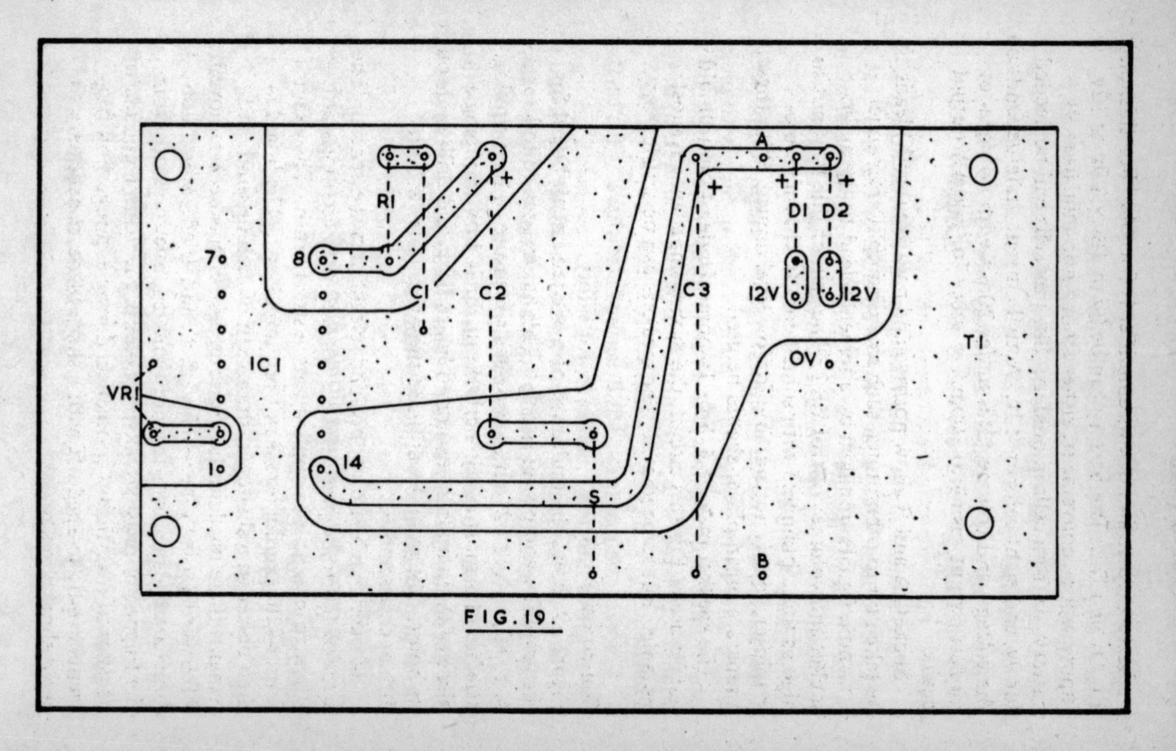

FIG. 19.

Holes for T1 should suit the fixing lugs. These bolts are long enough to allow mounting the board in its case, with clearance obtained by extra nuts or spacers. The secondary centre tap is connected to the negative line, and each 12v tag to one rectifier.

The IC must be placed the right way (see Figure 1). Also solder on leads for VR1, speaker S, and at A and B. Check that all joints are sound, and cut off excess leads.

A metal case is most suitable, though not essential. Fit a jack socket near VR1, which should be close to IC1 end of the board. With an insulated case, connect the control metal bush to negative. Fit a non-reversible low voltage socket, for connections A and B.

By having the speaker in the case, power supply, amplifier and speaker form a single unit, and it is only necessary to plug in a suitable audio signal. If a separate speaker were used, it could be of larger size, but must be fitted in a cabinet, or have a baffle.

Connect the mains cord as described elsewhere. This earths the negative line and metal case, for safety.

Point A is the supply for preamplifiers or other items, and a series resistor will be present in such additional equipment (as in Figure 16). The extra unit will also have a capacitor from its positive line, to negative. This provides additional smoothing, and decouples the positive line against feedback. B provides a direct ground connection between units.

Components:- Mains IC Amplifier (Figure 18)

Semiconductors

IC1	LM380		
D1	1N4001	D2	1N4001

Resistors

R1	2.7Ω
VR1	50k Log. potentiometer with switch

Capacitors

C1	0.1μF	C2	470μF 16v
C3	2500μF 20v		

Miscellaneous
T1 250v/12–0–12v 250mA Transformer
8Ω loudspeaker about 3½ x 5in (90 x 127mm)
Board, sockets, case, etc.

741 IC

The 741 is an inexpensive and very useful integrated circuit amplifier. The leads of the metal 741 can be arranged to fit perforations of either 0.15in or 0.1in matrix boards, and the lead numbering here is for this type.

Negative is 4, and positive 7, and up to 36v may be used. Good results are obtained with lower voltages. Output from the IC is at 6, and is from a complementary pair of transistors. There are two inputs. That at 2 is the negative or inverting input, and can be employed for negative feedback. That at 3 is the positive, or non-inverting input. Leads 1 and 5 provide connections to internal offsets.

Audio Amplifier

An audio amplifier to boost headphone reception, or to use as a preamplifier, can be made with the circuit in Figure 20. Audio input is to the isolating capacitor C1,and 3 of IC1. R1 and R2, of equal value, set the operating point.

Output from 6 may go to high impedance phones, or a screened lead can feed the main amplifier from here. The IC amplifier, used alone, will be found excellent for such purposes as boosting the signals from a crystal diode or other simple receiver.

Negative feedback is arranged from output 6, to the inverting input 2, by means of R3. Increasing the value of R3 reduces feedback, while increasing the value of R4 increases feedback. C2 is a DC isolating capacitor, and is best not too large in value, to avoid a delayed effect when switching on.

Gain can be pre-set at various levels, by altering R3. The circuit can be used with a 9v battery, but operates well with as little as 3v. Over 9v may be used, if power is drawn from a main amplifier.

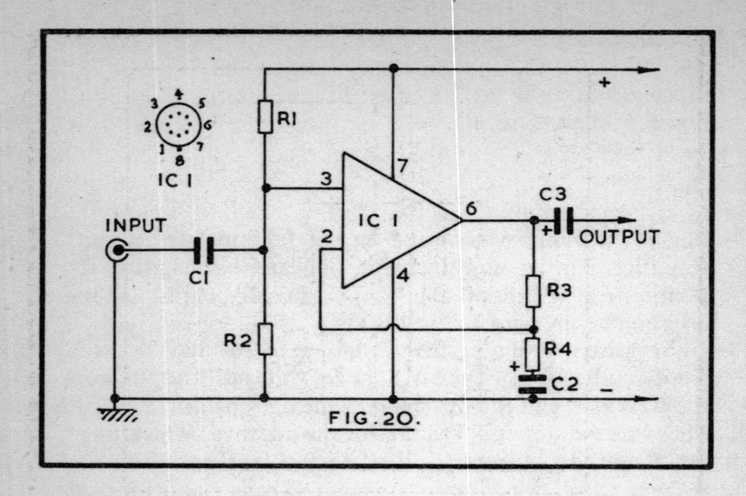

FIG. 20.

Construction
Figure 21 shows assembly on the board. Leads 1, 5 and 8 are not used but should not touch other metal items. The amplifier is easily tested by connecting a crystal diode radio or other suitable input, and phones or amplifier.

For many purposes it can be used in this form, by fitting the board in the receiver or amplifier. Alternatively, other features can be added, so as to increase its scope.

Modifications
In Figure 22 at A, variable negative feedback is used for gain control. R3 is removed, and the 5.6k resistor and 500k potentiometer are used to replace it. R4 and C2 remain as originally. Gain is maximum with the full resistance in circuit, and falls as the potentiometer is rotated to increase feedback. A 470k component is equally suitable.

At B, selective feedback is used for tone control. R3 is 100k, and in parallel with the 220k potentiometer and 10nF capacitor. Again, R4 and C2 remain as shown earlier. This circuit provides adjustable feedback of higher frequencies.

Individual controls of this kind will most likely be required if the IC is used for a separate preamplifier. For battery

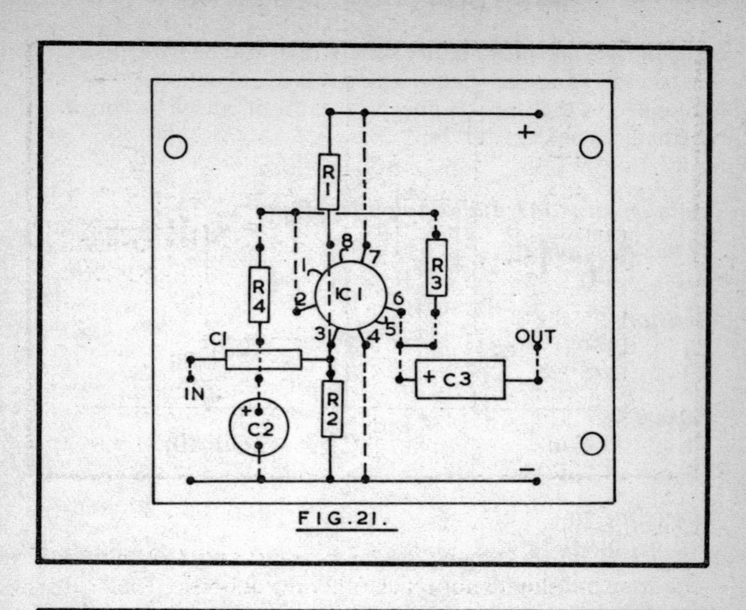

FIG.21.

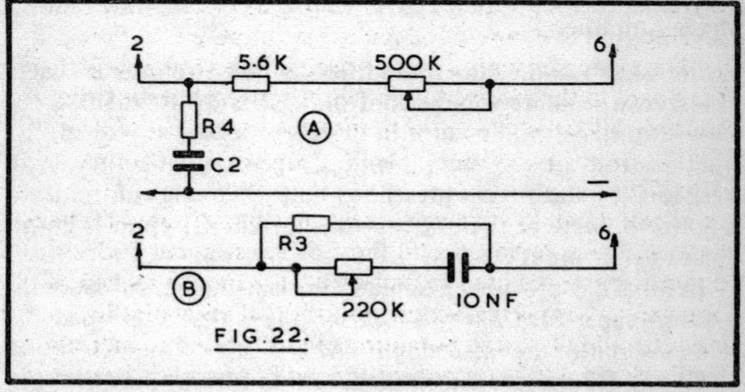

FIG.22.

operated equipment, or where there is relatively low gain, satisfactory results are possible with unshielded audio leads. These should not be longer than reasonably convenient. But for most sensitive amplifiers, screened audio leads will be required, to avoid picking up mains hum, or introducing instability. The preamplifier should then also be fitted in a metal case, for screening purposes, and this should be grounded to the negative line. This must of course be correctly earthed, for safety.

Components:- 741 Audio Amplifier (Figure 20)

Semiconductors

IC1	741		

Resistors

R1	100k	R2	100k
R3	56k	R4	1k

Capacitors

C1	0.47μF	C2	10μF 10v
C3	4.7μF 10v		

Miscellaneous

Board, Sockets, etc.

Mixer-Amplifier

Figure 23 shows the circuit modified to take two inputs. Each has its own separate volume control, VR1 and VR2. These allow two signals to be faded in or out, or mixed as wished, whether from a radio tuner, pick-up, tape or other source. Resistors R5 and R6 are present so that when one control is set at zero, input at the other is not shorted. C1 couples both signals to the input 3.

Feedback is provided as already explained. Output from the isolating capacitor C3 is to a jack socket, so that a screened lead may run to the main amplifier.

C4 and R7 provide decoupling, so that current may be drawn from the main amplifier power circuit. It is most convenient to have a non-reversible socket on the amplifier, so that a plug from the mixer-amplifier can be inserted. It is then only necessary to plug this in, and the audio lead.

With 12v available at IC1, current taken will be about 10mA. If necessary, R7 can be modified with this in mind. R7 will

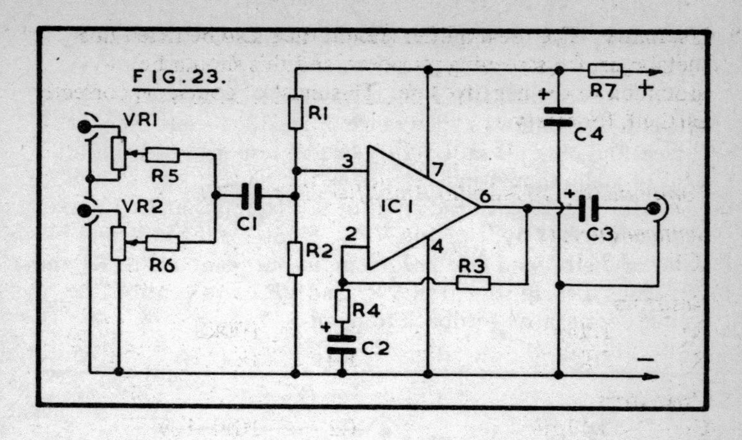

drop approximately 1v for each 100 ohms, but IC1 can receive quite a high range of voltages.

Components:- Mixer-Amplifier (Figure 23)

Semiconductors

IC1	741		

Resistors

R1	100k	R2	100k
R3	56k	R4	1k
R5	220k	R6	220k
R7	220Ω (12–18v supply)		
VR1	1M log. potentiometer	VR2	1M Log. potentiometer

Capacitors

C1	0.47μF	C2	10μF 10v
C3	4.7μF 15v	C4	220μF 15v/25v

Miscellaneous

Board, Case, Sockets, etc.

Feedback Treble-Bass Control

By using negative feedback which is sensitive to changes in frequency, a very effective tone control can be made. A single

IC circuit with both treble and bass controls is shown in Figure 24.

Treble boost and treble cut are available, and bass boost and bass cut. This allows a wide range of results, to suit the user's choice, and obtain a satisfactory overall response with all sorts of audio inputs and amplifiers.

Audio signals reach the input of IC1 from R3 and C3, levels being controlled by VR1 and VR2. Point 2 is the inverting input of the IC, and feedback from output point 6 is to R1 and C2. Thus the adjustment of VR1 and VR2 also controls the amount of negative feedback to point 2.

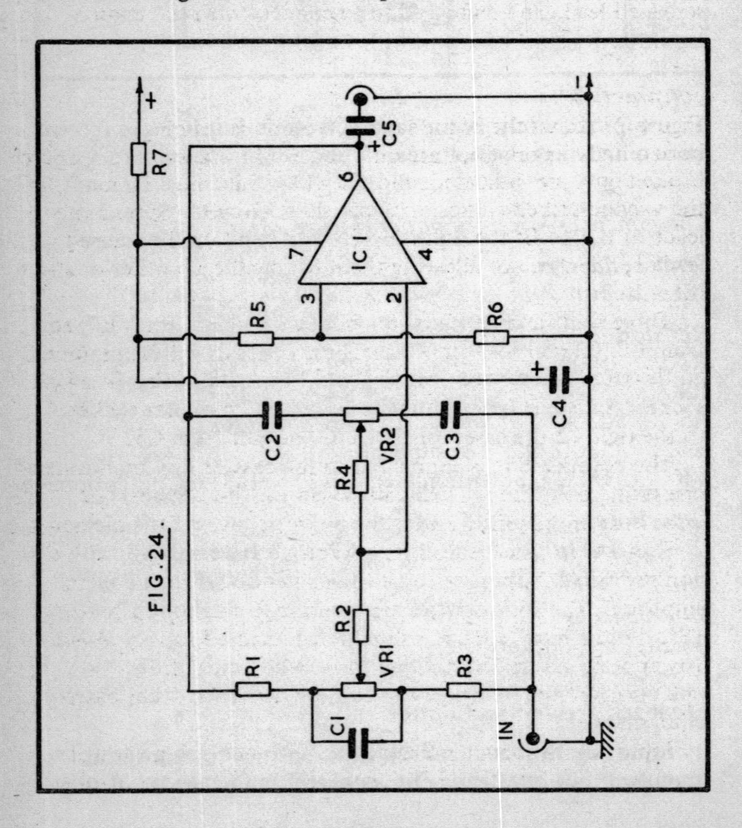

FIG. 24.

VR1 is the bass control, with C1 across it. VR2 is for treble, coupling and feedback being through C3 and C2, which fall in reactance as frequency rises. Such a tone control is much more effective than one operating to cut higher frequencies only, or attenuating a range of frequencies without feedback. When VR1 and VR2 are set in the middle of their travel, a "flat" response is obtained.

R5 and R6 supply input point 3. R7, with C4, decouples the tone control supply at 7 from the positive line, so that it can receive current from a main amplifier, without introducing hum or instability. C5 couples to an output socket, so that a screened lead can be plugged in to connect with the main amplifier.

Construction

Figure 25 shows the board layout, with foils running horizontally. Foil strips are only required for conductors where broken lines are indicated, and elsewhere cuts must be made, in the way described earlier, to avoid short circuits. Spread the leads of IC1 to fit the holes shown, and bend up the unused leads 1, 5 and 8, not allowing them to touch each other or any other items.

Three thin flexible leads run from C1 and R2, for VR1; and from C2, C3 and R4, for VR2. These controls will be mounted on the front of the case, which should be metal. Also fit a jack socket to the left here, for audio input, and a similar socket to the right of the case, for the audio output from C5.

The negative line is connected to the case or box by means of one fixing bolt, the tag being provided for this. Spacers or extra nuts are required under the board to give a little clearance.

Run a twin flexible lead from positive and negative, with a non-reversible 2-pin plug to fit the power outlet of the main amplifier. The tone control can operate from about 12v to 18v. For high voltages, R7 should be increased so that about 16v appears across C4. C4 and R7 can be omitted, for operation from a 9v battery in conjunction with other battery equipment.

Input can be from a preamplifier, or tuner giving enough audio without boosting. The unit feeding the tone control

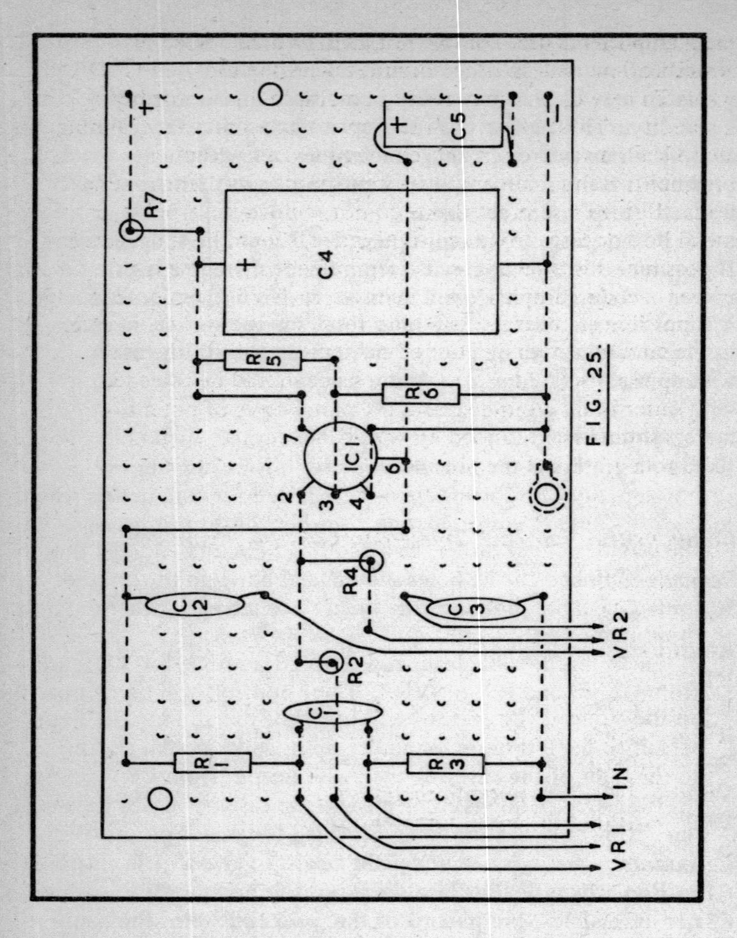

FIG. 25.

should have an isolating capacitor in its output circuit. If this is not present, a capacitor of 2.2μF or so can be included at the input socket of the tone control, for DC isolation. C5 provides isolation of output circuits.

As with preamplifiers and other units which may be used to feed a main amplifier, complete screening is not always necessary. This is most likely to be so when a battery operated main amplifier is used, or one not giving a high degree of amplification. But in other circumstances the use of an insulated case or box may result in mains hum, or possibly instability. These effects can be kept to a minimum by running audio leads away from mains connections, and keeping preamplifier and main amplifier input leads away from mains leads. But if these precautions do not remove the trouble, a metal box or case, grounded to negative, should be substituted. If possible, this type of case is recommended where a high power or mains amplifier will be used, to avoid trouble. But it is possible to secure satisfactory results without a metal case, in the circumstances mentioned, so some constructors may wish to use insulated cases. Audio leads should be screened, with outer braid grounded, and the metal cases of potentiometers should be grounded by wire connections, when not fixed to a grounded metal panel.

Components:- Feedback Treble-Bass Control (Figure 24)

Semiconductors

IC1	741		

Resistors

R1	4.7k	R2	39k
R3	4.7k	R4	5.6k
R5	100k	R6	100k
R7	680Ω		
VR1	100k Lin. potentiometer		
VR2	100k Lin. potentiometer		

Capacitors

C1	47nF	C2	2.2nF
C3	2.2nF	C4	220μF 16v
C5	4.7μF 16v		

Miscellaneous

Board, Sockets, Case, etc.

LM386 Amplifier

The LM386 is very useful for small, battery operated equipment, as it is very economical in current required, and will function satisfactorily from a low voltage. Typically, it can be used with a supply of 3v to 9v, though the lower voltage limit depends somewhat on the actual amplifier. Current drain is about 2mA to 20mA, so is easily within the capacity of a small battery. Over 12v should not be used.

Very few external components are required, Figure 26. VR1 is the volume control, and input can be from a tuner, pre-amplifier, or earlier stage. Where an external audio circuit is used, the lead will be screened in the usual way, with the outer braid taken to the negative line.

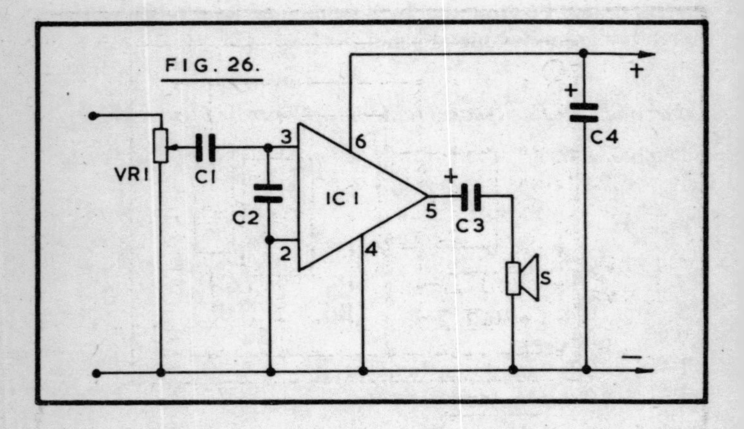

The speaker S is 8 ohm impedance. For miniature equipment, a 2¼in to 3½in (58–90mm) unit will be satisfactory. Otherwise, a larger speaker can be used, with some improvement in reproduction. The speaker should have a cabinet or baffle, or a loss of volume and low frequencies is likely. The IC is intended for small equipment, and power output is up to about 250mW. This is easily sufficient for average domestic purposes.

Construction

Figure 27 is a layout for the amplifier. It is necessary to use 0.1in matrix board, so that IC1 will fit. Broken lines show essential circuits completed by the foils below the board. Elsewhere, the foils must be cut to avoid short circuits. Such breaks are needed between 1 and 8, 2 and 7, 3 and 6, and 4 and 5 under IC1, under C3 and C4, at S, and between C1 and C2 and the negative line. They can be made with a sharply pointed tool, with a spot cutter, or with a few light turns with a sharp drill. Examine the board to make sure the foils are completely cut, and that fragments of metal do not touch the adjoining foils. Cut away the foil around the fixing holes, or cut the strips near, so that no trouble arises here.

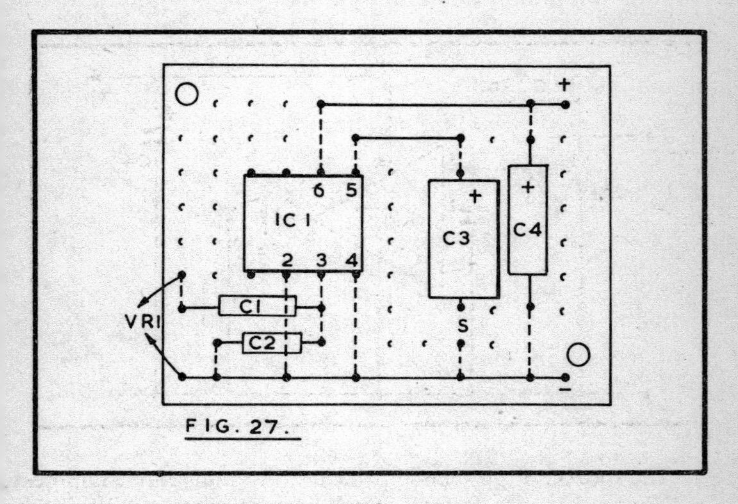

FIG. 27.

It will be found that output and gain is very well maintained with reduced voltage. The loudspeaker leads from S should run away clear of wires to the volume control.

If a 6v supply will be used and minimum size is wanted, the capacitors can be 6v or 6.4v. Capacitors of higher voltage rating can be fitted when size is not important.

Components:- LM386 Amplifier (Figure 26)

Semiconductors

IC1	LM386

Resistors

VR1	10k Log. potentiometer

Capacitors

C1	0.1μF	C2	470pF
C3	470μF 10v	C4	100μF 10v

Miscellaneous

Board, 8Ω Loudspeaker, 8 pin DIL holder, etc.

CHAPTER 4

MISCELLANEOUS PROJECTS

IC Receiver Source

This unit is a source of radio signals for a receiver, over the range approximately 1500 – 600kHz, or 200 – 500 metres medium wave. If it is used in conjunction with a receiver which normally operates from an external aerial, the latter is no longer required for reception in this band. It can also be employed with receivers having their own internal ferrite rod aerial, and can then allow the reception of stations which are not normally heard, or which are at too low a signal level for satisfactory reception.

The circuit uses the CA3028A integrated circuit radio frequency amplifier, Figure 28. Band coverage is determined by the ferrite rod winding L1, and tuning capacitor VC1. A commercially made MW aerial is suggested for use here, and those sold for the home constructor will usually be for a ganged

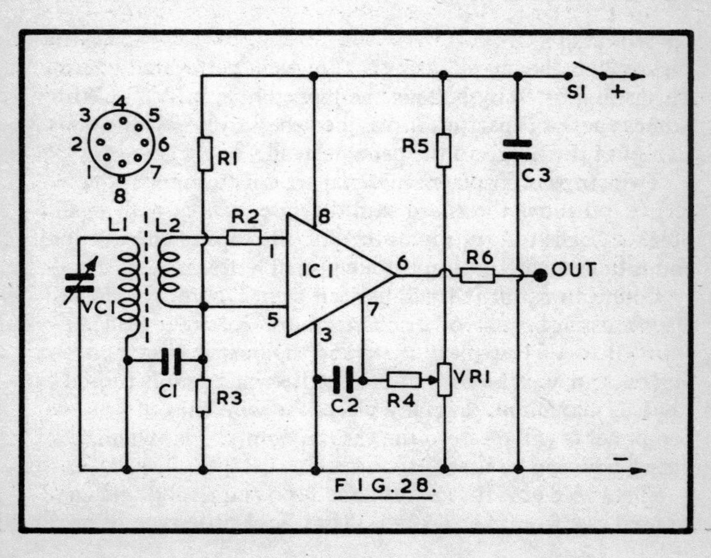

FIG.28.

capacitor with 208pF aerial section. VC1 can then be 200pF, or the 208pF section of the usual 208/176pF superhet type ganged capacitor. Larger values may be used, but the wanted range will then only occupy part of the tuning range. The capacitor is best of air spaced type.

The coupling winding L2 is five to seven turns, wound on the earthed end of L1. A suitable winding will be present on most coils, or can easily be arranged from thin insulated wire. R2 is to help preserve stability, and 10 ohm can be fitted here. Or lower values may be tried, or R2 may be omitted experimentally.

VR1 is a manual gain control. Without some means of reducing amplification, there is likelihood of overloading the receiver when strong transmissions are tuned in.

R6 has a similar function to R2, and the same notes apply. Current can be drawn from a small 9v battery, though 12v can be used. This could be from cells in the usual type of holders.

Various means of coupling the output to the receiver can be used, as described. These depend on the receiver details.

Construction

Figure 29 is the layout, using 0.15in plain perforated board. This is supported by bolts in the tapped holes of VC1. With other types of capacitor, it may be necessary to use brackets to mount the board to the panel instead.

Two strips of insulated material are cut to support the ferrite rod above the board, and they are secured with small brackets. The rod rests in notches in the tops of these strips, and is bound with cotton through small holes.

Connections for IC1 will be seen from Figure 28. Unused leads must not touch other connections. Keep the lead from L1 to VC1 up clear of R6 and the output circuit, as feedback may cause oscillation to arise when gain is turned towards maximum. In many places the wire ends of components will reach to the various points. Elsewhere, 22swg wire can be used.

The whole unit fits an insulated box, and the box lid can be used as a front panel. With the rather brittle type of box, drill

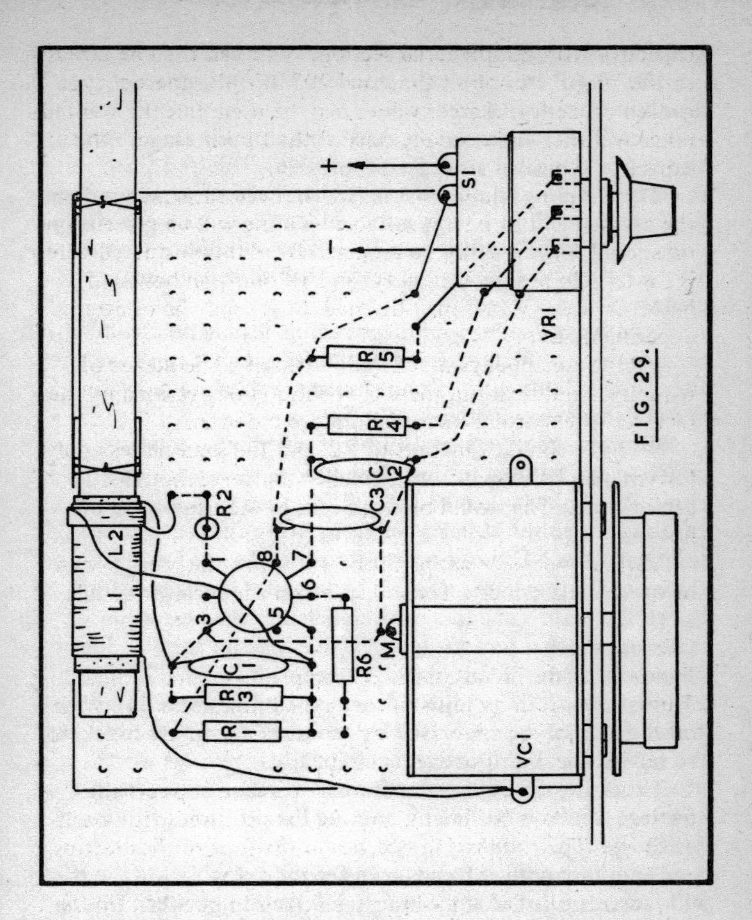

FIG. 29.

must be sharp, and used with only moderate pressure, or the material may break. A clear box of this type can be painted inside, and this will give a uniform external appearance.

Using the Unit

Means have to be taken to couple output at R6 to the input or aerial of the receiver. At the same time, the circuit from R6 to positive line has to be completed. For maximum possible results, the method of coupling should provide best transfer of RF energy to the receiver, while not causing instability or oscillation.

Where the receiver has an aerial coupling winding, this can be tried. Connect it from R6 to positive. The lead from R6, in particular, should not be longer than necessary. If required to avoid instability, a short screened lead may be used. The outer braid is the earth return, and inner conductor runs from R6. RF type co-axial lead is better than screened audio type cable.

If there is no aerial coupling winding, it may be possible to arrange one on the ferrite rod of the receiver. It can have about fifty turns of thin wire. Connections are then made to the ends of this new winding.

With some receivers, and those of older type, capacity coupling can be used to the aerial socket. A resistor or small cored RF choke is then connected from R6 to positive, and a capacitor of about 100pF goes from R6 to the receiver aerial terminal. The RF choke must be clear of L1, or feedback causes oscillation. It is possible to substitute a resistor for the choke, with some loss of overall gain. The best value depends on other factors, but 470 ohm can be used, and can be modified later if this improves reception.

In all cases VC1 is adjusted to peak up the wanted signal and the dial can be calibrated by reference to the receiver. As the ferrite aerial is directive, occasionally it may be worth while rotating the equipment for best results. Exact band coverage can be modified by altering the position of L1 on the ferrite rod.

Where the receiver has its own ferrite rod, position the unit and receiver to avoid unnecessary interaction between this and the circuit L1. A symptom of stray feedback is oscillation arising when VR1 is turned up for maximum gain, and also when VC1 is set for best volume from the required station.

Components:- IC Receiver Source (Figure 28)

Semiconductors
IC1 CA3028A

Resistors

R1	1.2k	R2	10Ω
R3	2.2k	R4	270Ω
R5	3.3k	R6	27Ω
VR1	10k Lin. potentiometer with switch (S1)		

Capacitors

C1	0.1μF	C2	0.1μF
C3	0.1μF		
VC1	208pF (or to suit L1)		

Miscellaneous

L1 MW ferrite aerial
Board, Case, etc.

The 555

Numerous circuits can be made around the 555 IC, which is intended for oscillators and timers. A basic circuit for using this IC is shown in Figure 30.

For operating, about 4.5v to 15v may be used, and sometimes this can be provided by associated equipment. Otherwise, a 9v battery can be fitted.

When the circuit is switched on, C1 is discharged, and 2 and 6 are negative. C1 charges through VR1, R1 and R2. When a certain potential is reached, C1 is discharged by the IC, and a pulse arises at point 3. The process is then repeated.

Smooth control over the frequency at which this happens can be obtained by means of VR1. Large changes in frequency can be arranged by selecting suitable values for C1. In this way, an exceedingly wide range can be covered. The IC can thus be used for audio oscillators, tracers and timers. Normally, point 3 is high, and it pulses low when C1 discharges. So the IC can also be used to provide an input for digital circuits.

Figure 31 shows the IC assembled on a board, and R2 can generally be 2.2k. C2 may be 0.1μF. The values to be fitted for C1, VR1 and R1 will depend on the purpose in view.

Increasing the value of C1 lowers the frequency. An increase in value at VR1 or R1 also lowers frequency. R1 is included so that some resistance remains in circuit, with VR1 set to

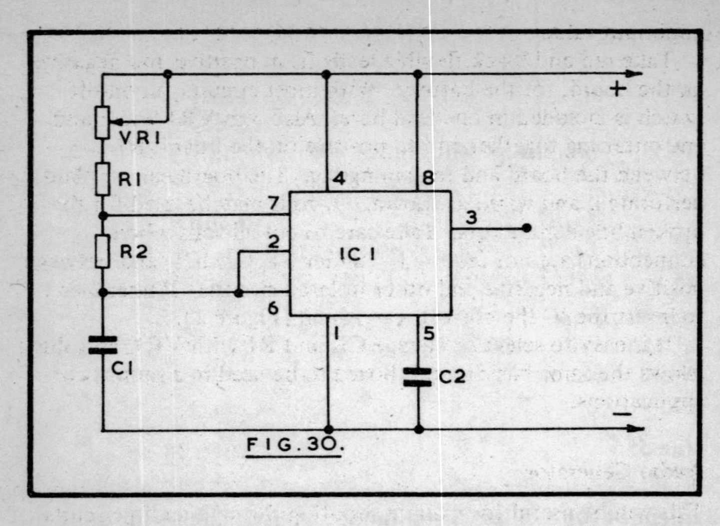

FIG.30.

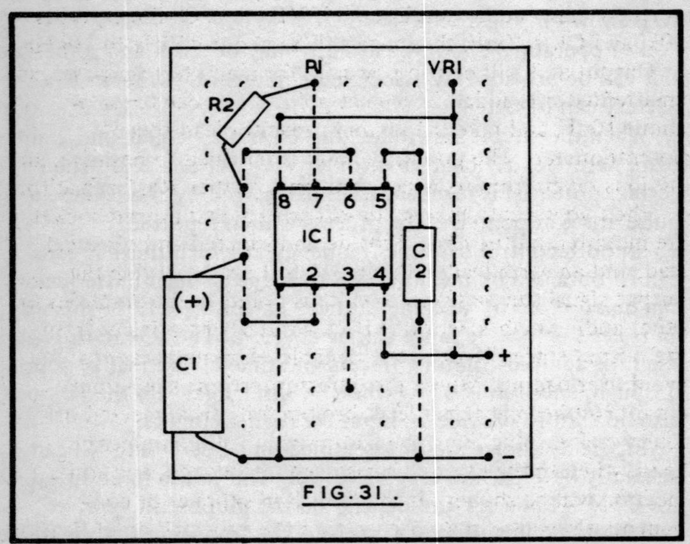

FIG.31.

minimum value.

Take red and black flexible leads from positive and negative on the board, for the battery. With most circuits, an on-off switch is included in one lead here. Also join VR1 wiper and one outer tag together and to positive on the board. R1 is between the board and remaining tag. The board can be plain perforated, and wired as shown. Or foils may be used for the broken line conductors. Take care to cut all foils where connections are not required – under C1, C2, IC1, and between positive and negative and other isolated circuits. Remember to insert the IC the correct way round (Figure 1).

It is easy to select or change C1, and R1 with VR1, and this allows the same basic circuit board to be used in a number of applications.

Audio Generator

This will be useful for trouble shooting through audio circuits in receivers, or audio amplifiers. If VR1 is 250k linear, R1 is 10k, and C1 is 10nF, the range will be about 950Hz to 10kHz.

Output at 3 will be too powerful for many test purposes, so an attenuator is added. Connect point 3 to a capacitor of about 10nF, and take the second capacitor lead to a 5k potentiometer. The potemtiometer is returned to negative, and audio is taken from its wiper through a further 10nF capacitor.

The two potentiometers, on-off switch, and output socket are mounted on the case. Provide an insulated and screened lead, and an earth lead with clip. A test prod carrying the output signal can be taken to various points in the amplifier or other audio circuits, working backwards systematically from the output stage. Reduce the signal level, as necessary to avoid overloading. Where the defect interrupts the audio circuit (broken capacitor lead, broken foil, disconnected or faulty jack or plug, etc.) the audio signal will no longer be heard when the faulty items has been introduced, and so its location will be shown. If the fault is in supplies or components elsewhere in a stage, so that the expected amplification is not obtained, or signals cease, then individual testing of resistors and other items in the defective part of the circuit will be needed.

Do not work on AC/DC mains equipment, or any apparatus having a live chassis, or high voltages.

The audio generator will also serve as audio source for the quick checking of audio amplifiers. The output attenuator described will normally have to be set to a low level.

Morse Oscillator

Output is sufficient for a loudspeaker, or can be reduced for headphones. The output coupling capacitor is best increased in value to about 10μF. The capacitor from potentiometer to output lead is not required, as the speaker can be connected from here to the key, which is returned to negative. The oscillator is thus left running, and code is produced by completing the speaker circuit with the Morse key.

A speaker of about 8 ohm or 16 ohm impedance can be used. Other impedances will be satisfactory. VR1 is set to give a suitable pitch, and volume is adjusted with the second potentiometer. As current is drawn in pulses, a 100μF capacitor can be connected across the supply at the board, and can increase audio output.

Morse is used by Scouts, Radio Amateurs, and others. The code will be found in "Electronic Projects for Beginners", Book Number BP48, also published by Bernard Babani (publishing) Ltd.

For very long periods of running, one of the power supplies described may be used instead of the battery.

Digital Pulser

With TTL digital ICs of the 7400 types, a 5v positive supply line will be available, and this is used for the 555. Provide leads and small clips, so that positive and negative can be attached to any convenient circuit points, to provide current for the 555.

A lead with a series resistor of about 150 ohm is taken from point 3, and equipped with a clip or prod. This can be applied to various points in the digital circuit, such as input to tens, hundreds, and later dividers; or to seconds, minutes and hours dividers. By this means, the counters and their displays can be stepped ahead at any required rate.

If VR1 is a 500k linear potentiometer, and R1 is 4.7k, C1 can be 1μF, for approximately 3Hz to 160Hz. For a much faster rate, for rapid testing of a string of dividers, C1 can be reduced to 0.022μF, this providing about 200Hz to 6,000Hz. Where a very slow pulse is required, so that individual circuits can be checked, C1 can be increased to 33μF, for a range of 3Hz to one pulse per 20 seconds.

With an electrolytic or polarised capacitor, connect as in Figure 31. Leakage in the capacitor will reduce the rate, especially for large values at VR1. C1 should be of good quality, or may be tantalum.

Each pulse at 3 drops low, returning almost instantly to high. This is suitable for most purposes. Where investigation of circuits requires "high" and "low" outputs of approximately equal period, disconnect tag 7. Also take VR1 to 3, instead of to the positive line. Output time for each state is then approximately equal. As example, 10 seconds high, and 10 seconds low, at the 3 pulses per minute rate.

Circuits will be seen later, wired as in Figure 31, where the IC is used as input for dividers.

Interval Timer

With a 12v supply, and VR1 increased to 1 megohm, with 47k for R1 and 47μF for C1, a delay of approximately 3 to 60 seconds will be obtained. Point 3 can be used to control a relay directly, or to provide base current for a switching transistor, which in turn controls the relay. The second method relieves the 555 IC of the relay current, and allows the use of a low resistance relay, if wished.

Figure 32 shows a circuit which is suitable for a 12v 100 ohm or similar relay. When the push-switch S is pressed, current reaches the relay and Tr1, and the 555 IC. The latter is positive at 3, and current through R1 to Tr1 base allows collector current to energise the relay, so that the contacts close. The circuit thus remains on, when S is released. After

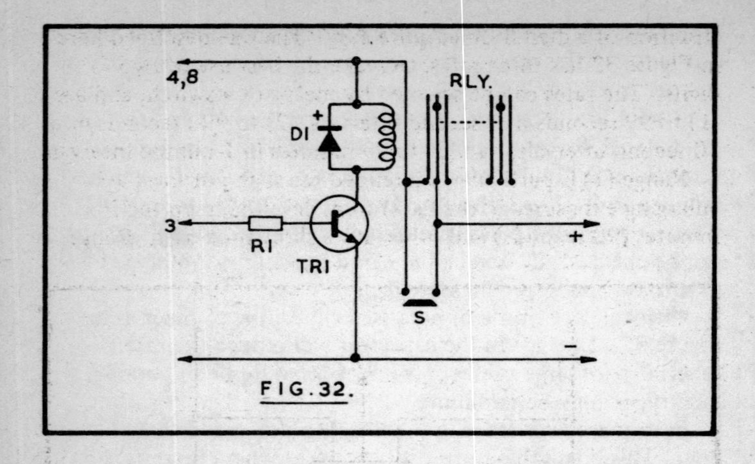

FIG.32.

the elapsed interval, which has been set by VR1, the base of Tr1 is moved negative, and the relay is released. This interrupts the circuit to the 555, so that the sequence is not repeated until S is pressed again.

The second set of relay contacts will be used to control the required equipment. This may be an indicator or enlarger lamp, or any device where repetition of timing is wanted when S is pressed.

A mains voltage lamp can be controlled by the relay, provided this is of the type with contacts for the purpose. Mains wiring must be separated from other circuits, for safety.

D1 suppresses back emf when the circuit is broken. The 1N4002 is suitable here. R1 can be 6.8k, and Tr1 a BC108, 2N3706, or other transistor suitable for the relay current required. Collector current in the "on" condition can be adjusted by changing the value of R1, but if current is too low the relay will not operate, while excess current will heat Tr1, so if necessary check that the rating for the latter is not exceeded.

Digital Timer

Readily available digital integrated circuits make the construction of a digital timer quite easy. The one described here in Figure 33 has three rates, to make the best use of two digits. The rates can be selected by means of a switch, and are (1) to 99 seconds in 1 second intervals, (2) to 990 seconds in 10 second intervals, and (3) to 99 minutes in 1 minute intervals.

Range (1) is particularly intended for such purposes as enlarging exposures in the darkroom, developing up to 1½ minutes (90 seconds) and other quite short processes. Range

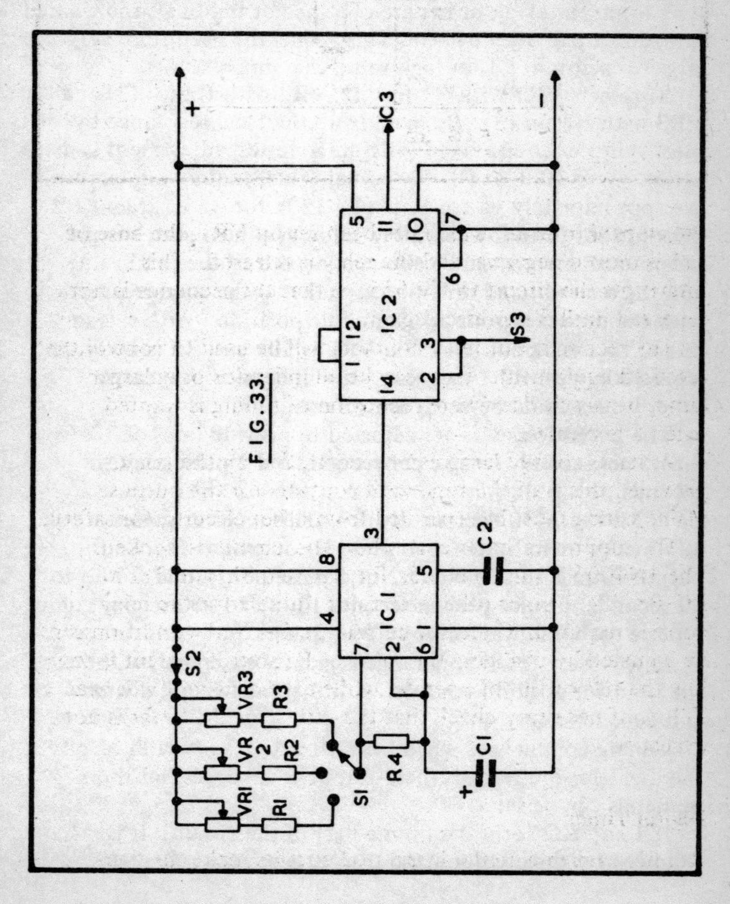

FIG.33.

(2) is convenient when a somewhat longer period is wanted and timing needs to be close – such as tank development. Range (3) is for long periods, such as fixing or washing, where timing to 1 minute is easily adequate, and various other domestic, game, sport and general activities will also fall within one of the ranges.

The 555 timer IC is used as the clock, Figure 33. Because of the difficulties which sometimes arise when arranging for a very long timing interval, pulses are obtained at ten times the wanted rate, and are divided by ten. This avoids the need for a very large capacitor at C1, or high value charging resistors.

S1 selects VR1 with R1 for (1), VR2 with R2 for (2) and VR3 with R3 for (3). By making up the total resistance by means of pre-set and fixed resistors in series, adjustment is made easier. With 10μF at C1, the actual values required in circuit are approximately 9k for Rate (1), 120k for Rate (2) and 1.2 megohm for Rate (3). To allow enough adjustment, the pre-set potentiometers can be about one-third of the total value, and the related fixed resistor can make up the total required, with the pre-set at roughly its middle position. With this in mind, VR1 is 5k and R1 6.8k, VR2 is 50k and R2 is 100k, and VR3 500k with R3 1 megohm. It will be seen that many other values could be used, as explained.

The potentiometers are adjusted to provide rates of 10 pulses per second, 1 pulse per second, and 1 pulse each six seconds.

IC2 is the 7490 decimal divider. This receives pulses at 14, divides them by ten, and provides output at 11. So the output from IC2 is at the rate of 1 pulse per second for up to 99 seconds, 1 pulse per ten seconds for up to 990 seconds, and 1 pulse each minute, for up to 99 minutes.

In this way, the timing range is greatly extended, without the need for several numerals, with their associated dividers and decoders.

Timing commences when S2 is closed. S3 operates on all the dividers, and returns their output to zero, so that the numerals can be set at 00.

IC1 and IC2 form the timing part of the circuit. It is followed by the counting and display section.

Counter and Display

Figure 34 is the circuit of this part of the timer. Pulses from 11 of IC2 are taken to 14 of IC3. This IC is a binary coded decimal divider, and provides a binary coded output along the points 11, 8, 9, 1 and 12. The binary code for the count received at 14 goes to 6, 2, 1 and 7 of IC4.

IC4 is a decoder-driver. Its purpose is to receive the binary input and to decode it in such a way as to provide output at the appropriate circuit points 9 to 15. These points are connected through current-limiting resistors to the numbered pins of the 7-segment LED numerals, 15 to 2, 14 to 11, and so on. The decoder-driver is thus able to illuminate various sectors of the

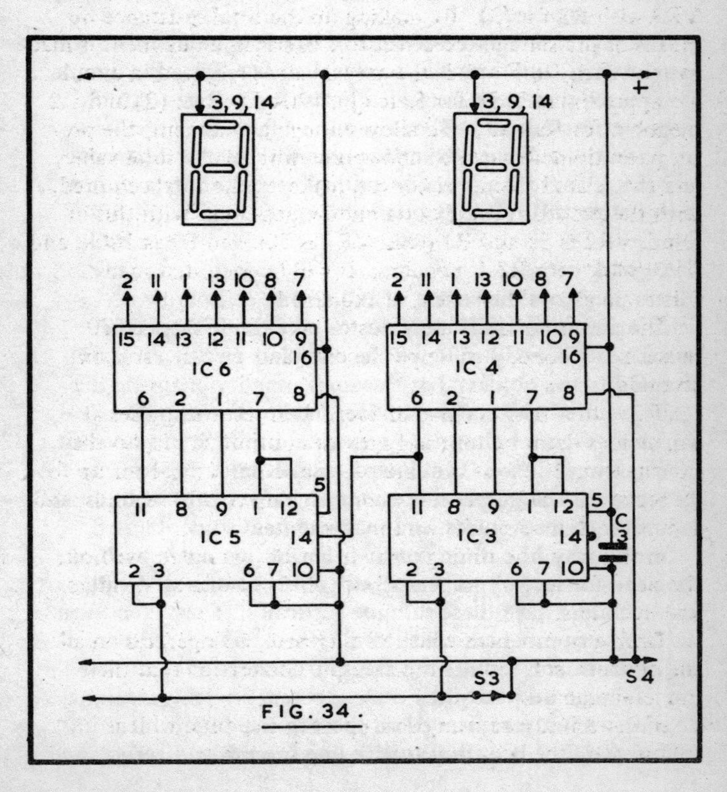

FIG. 34.

numeral, to produce the required figure.

At the pulse after 9, a pulse passes from 11 of IC3 to 14 of IC5, which has its own decoder-driver and numeral. So the four ICs and two numerals count up to a total of 99.

Positive to IC3 and IC5 is 5, and negative 10. With IC4 and IC6 positive is 16 and negative 8. The numerals have 3, 9 and 14 as positive (some numerals do not actually require all these pins to be taken to the positive line).

The on-off switch is placed in the negative line (S4). When power is switched on, the states of IC3 and IC5 may result in a random, unwanted display. This is cleared, and the numerals are set to zero, by opening S3. During normal working, S3 is closed.

The digital integrated circuits IC2 to IC6 are intended for 5v operation, and normally receive 4.75v to 5.25v maximum. This means that most satisfactory running will be from the 5v power pack described, or from the adjustable pack, set for 5v output. Some ICs of this type will be found to work satisfactorily from a 4½v dry battery, while others may not. A 6v battery **MUST NOT** be used.

Construction

Figure 35 shows the layout of IC1 and IC2. Board with perforations at 0.1in is required, to take the ICs, and the board is sufficiently wide for the complete circuit. It is not essential to use holders, but this does simplify changing the ICs if required, or to allow testing one not known to be working. Figure 1 shows the way round these items must fit.

When working on 0.1in board, a small soldering iron, as intended for the purpose, is essential. A large iron will tend to bridge adjacent joints, and may overheat some items.

Wiring may be carried out in either of two ways, each of which is satisfactory. For one, use the type of board with foil strips and have these running vertically. They then form the connections which are shown by broken lines, and are under the board. Solid lines are wire connections across the top of the board. Provided these are kept straight, insulated sleeving will only be needed on the negative line, which runs from 6 and 7 of IC2, the positive line from 5, and zero setting line from 2 and 3. It is helpful to use black, red and some other colour for these three circuits. All the current carrying leads need to be about 22swg. The other connections can be of thinner wire. Stout wire is awkward to deal with.

The second method is to use plain perforated board. Foils can be removed either by placing the whole board in etching solution, or by the use of a fine-grade power sander disc out of doors, if plain board is not available. Connections are then made throughout with wire. This can be about 28swg, for convenience in handling, except for power circuits (5, 10, etc.) where 22swg is recommended. Tinned copper wire will solder readily.

As an example of this method, solder the wire to 14 of IC2. Cut a few inches, and take up through the hole shown, and across the top of the board, down through the hole, and to 3 of IC1. Pull taut and solder on here, cutting off excess. By this means it will be found that wiring can be carried out quite rapidly. There will be no unwanted foils to cut, as when these are used for the underside conductors. The number of actual soldered joints is reduced.

Point A, from 7 of IC1 and R4, has a thin insulated flexible lead to take to S1. Points B, C and D will be wired to the other tags of this switch, which selects the range.

S2 is connected to leads which come from the positive line (5 of IC2) to the potentiometers. S3 is connected from the negative line to the circuit mentioned, which runs to 2 and 3 of IC2, IC3 and IC5.

The main on-off switch, S4, is included in the negative power line.

Figure 36 shows the left-hand part of the board, carrying the dividers, drivers and numerals. It is easiest to wire IC3 and IC5 at the same time. As example, 1 to 12 on each. Then 12 to 7 on IC4 and IC6. Then 6, 7 and 10 together, to negative, and to 8 of IC4 and IC6, and so on.

Remember that if foils are used, they must be cut where circuits are not shown. That is, under all ICs, between ICs where no circuits are required, and so on. This should be done systematically, and carefully checked as otherwise a fault may be introduced.

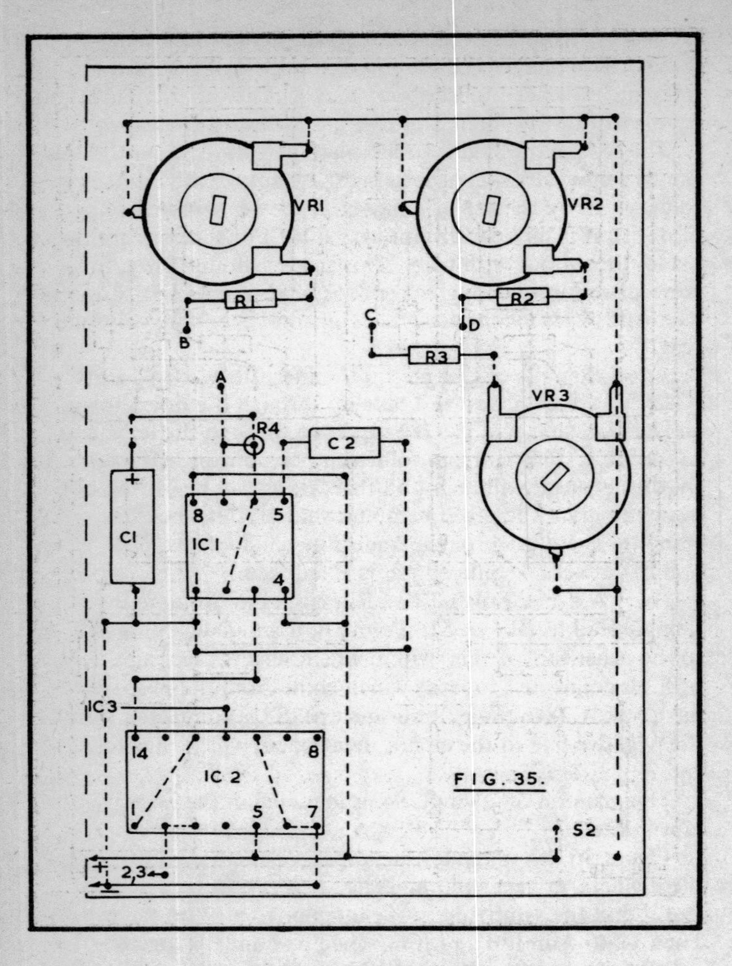

Each numeral sector has its own resistor. These need to be of miniature type, as they can then be most easily fitted. They can be 1/8 watt. Again, it is quicker to do both numerals together. As example, take a resistor and shape its leads to fit from the points driver 9 to numeral 7, solder on, and repeat exactly for the other numeral. Note that the numerals

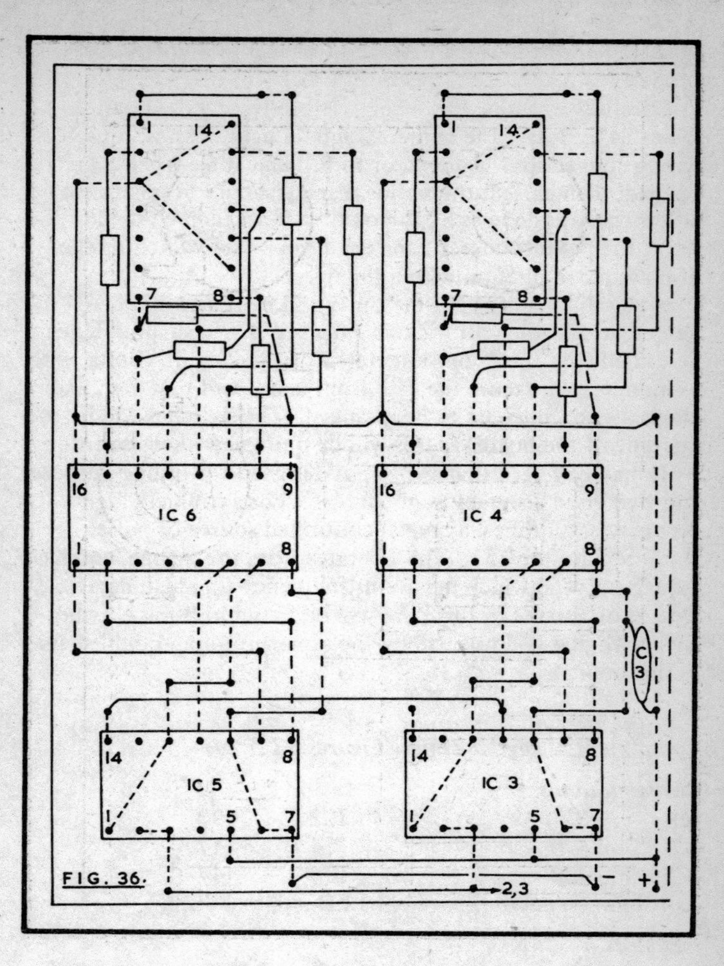

must be of common anode type (not common cathode). The size with 8mm figures is convenient. These will fit 14 pin DIL holders, if wished, and this helps raise them up from the board. The numerals have a mark indicating the top, and this must be uppermost, or the display cannot be correct.

A thin insulated wire runs from 11 of IC2 to 14 of IC3. A

similar connection is made from 11 of IC3 to 14 of IC5, as shown.

Carefully examine the wiring, before testing the board. When S4 is closed, the numerals should light. If not, look for wrong connections to them, or to IC4 and IC6. When S3 is opened, then closed, the timing display should begin at zero, and proceed so long as S2 is closed. S1 can then be used to select the potentiometers, and each can be set, watching the display in conjunction with a timepiece.

A suitable housing for the timer will place the numerals behind an aperture cut for this purpose. Three small holes can be drilled over the potentiometers, for adjustment. Bolts, with extra nuts, will secure the case front and board together, and allow the spacing here to be arranged. The switches will be to one side of the board, as this will lie quite near the case front.

It may be noted that the 555 is designed for timing purposes, and that good accuracy is obtainable. For extremely high accuracy at all times, a crystal controlled source of pulses could replace the 555. This operates from the 5v line, and has a quartz crystal which maintains frequency to a high degree. Details of circuits of this kind will be found in Book Number BP84 "Digital IC Projects" by the same author and publishers as this book.

Components:- Digital Timer (Figures 33 & 34)

Semiconductors

IC1	555	IC2	7490
IC3	7490	IC4	7447
IC5	7490	IC6	7447

2x Common anode 7-segment LED numeral display

Resistors

R1	6.8k	R2	100k
R3	1M	R4	2.2k
VR1	5k pre-set	VR2	50k pre-set
VR3	500k pre-set		

14x 270Ω ¼W for numeral displays

Capacitors

C1	10μF 6v	C2	0.1μF
C3	10nF		

Miscellaneous

S1	Single-pole 3-way switch
S2	On-off switch
S3	On-off switch
S4	On-off switch

DIL 8-pin holder
3x DIL 14-pin holders
2x DIL 16-pin holders
Board, Case, etc.

IC Metronome

A metronome will give the beat and guidance on rate for pop, classical and other music, and is very useful during practice. Figure 37 shows the circuit.

These values are arranged so that the rate, adjustable by VR1, is from about 45 to 184 beats per minute. Assemble the components on 0.1in perforated board, as described earlier,

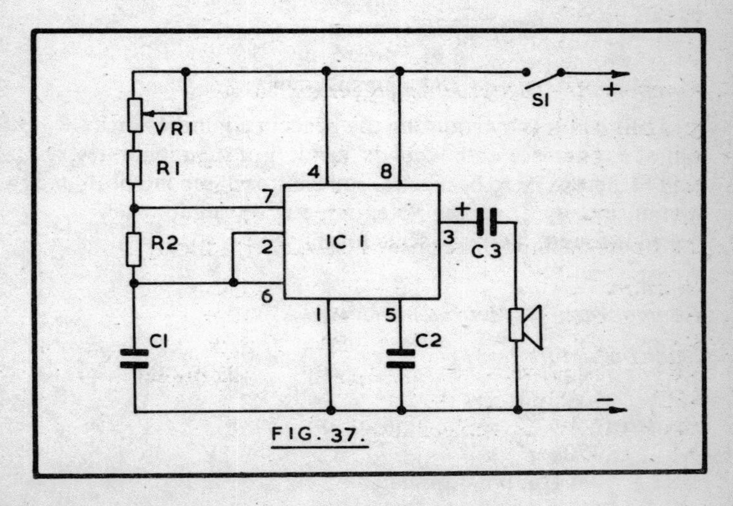

FIG. 37.

with leads for external connections.

Construction

Figure 38 shows a suitable case, constructed from thin wood. The back should be fixed by small screws only, so that the battery can be replaced. An opening is cut to suit the speaker cone, and thin fabric is glued behind this, before cementing the speaker unit in place. The bottom has four rubber feet, to avoid scratching polished furniture.

The front takes the slide switch S1, and rate control VR1. A separate switch is most suitable, so that VR1 can be left set when required. Fit a battery lead connector for 9v battery.

A scale is drawn for VR1, and fixed with adhesive, and a knob with pointer is put on the spindle. This scale is calibrated directly in rates and the appropriate musical terms, from the following list:-

46	largo
54	adagio
60	andante
80	moderato
100	allegretto
116	allegro
126	vivace
144	presto
184	prestissimo

Calibration is by counting the beats in a minute, with the aid of a timepiece with seconds' hand. For the rapid rates, it will be necessary to beat time, counting each second beat, and to multiply by 2. As the potentiometer setting for each speed is found, mark the scale for it.

Components:- IC Metronome (Figure 37)

Semiconductors

IC1 555

Resistors

R1 470k R2 2.2k

VR1 2M Lin. potentiometer

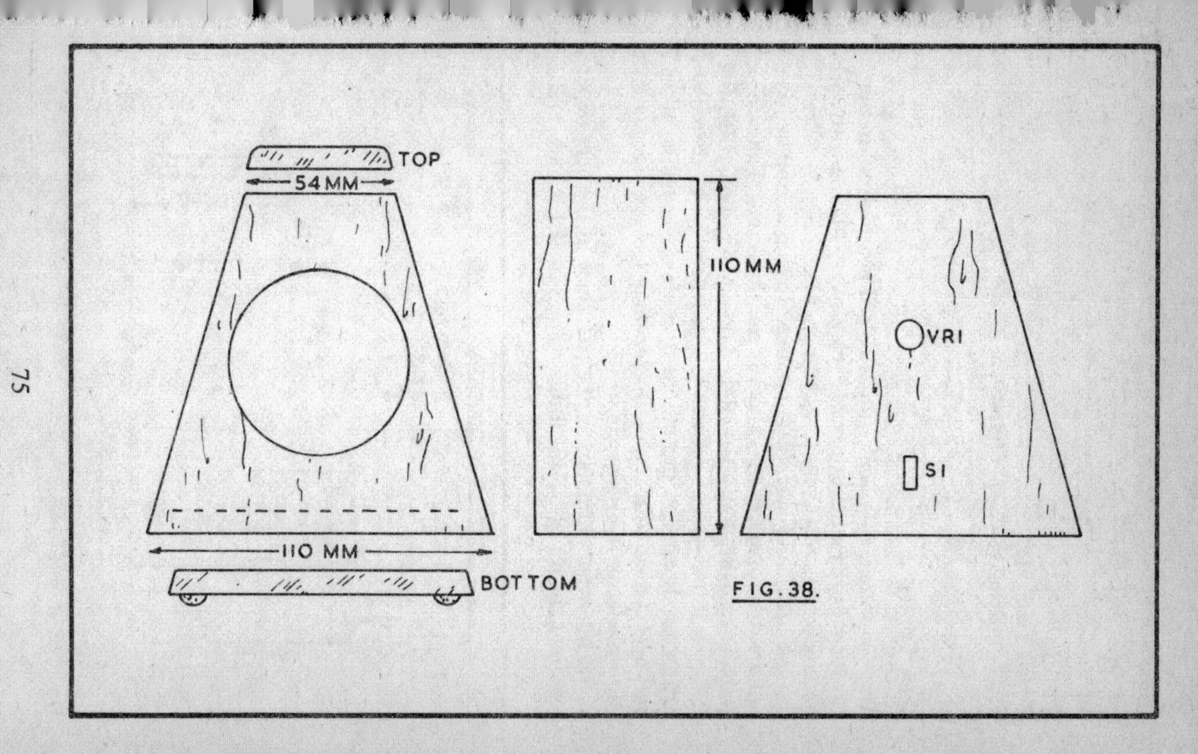

FIG. 38.

Capacitors

C1	1μF Paper	C2	0.1μF
C3	22μF 10v		

Miscellaneous

Board, Case, etc.
8Ω Loudspeaker 2½in (60mm)

Dark Trigger

The 741 integrated circuit amplifier will prove useful in the construction of a light operated relay. Very high sensitivity is obtained, with relatively few components.

The circuit in Figure 39 is arranged so that a drop in illumination operates the relay. It can thus control an emergency light, which is switched on when ordinary lighting fails. Or it may switch on a child's bedroom light, when daylight goes. It can also put on one or more house lights, as darkness comes, which householders may feel is a safeguard or convenience in their absence.

With normal illumination, the resistance of the light dependent resistor LDR is low, so the input point 3 of IC1 is high, and

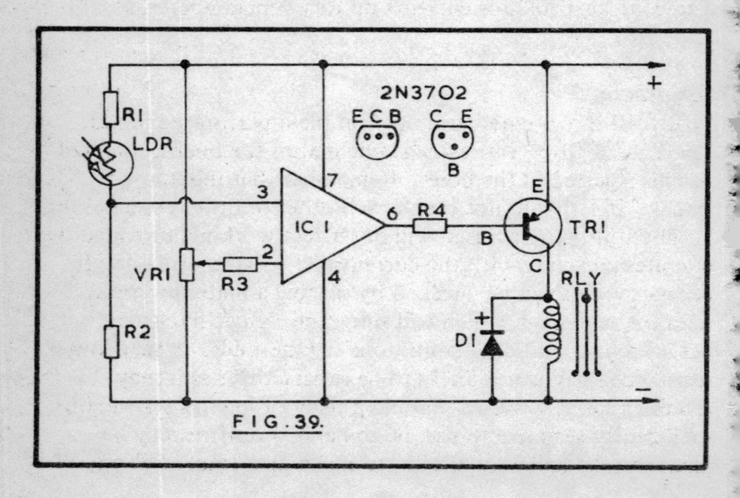

FIG.39.

so is the output point 6. Tr1 is a PNP transistor, so collector current is very small while the base is held positive. In these circumstances, negligible current passes through the relay.

When light reaching the LDR falls, its resistance increases, and 3 moves negative. Output 6 can now supply base current through R4 to Tr1, and collector current rises, operating the relay. With a 9v supply, "light" current through the relay was found to be nearly zero, rising to about 55mA under "dark" conditions. D1 is to suppress back emf.

VR1 allows adjustment of the inverting input potential of IC1, and thus the operating point, or sensitivity to illumination. This potentiometer is useful if operation is wanted at low levels of illumination, and small changes of lighting.

A 150 ohm relay was fitted. Collector current is to some extent controlled by the value of R4, as well as the gain of Tr1, voltage, and relay resistance. Provided Tr1 is used within its ratings, and the relay operates strongly, considerable modification to these items will be possible.

A relay with low voltage contacts is suitable for low voltage lamps and similar purposes, such as switching on a child's nightlight. When mains voltages are to be handled, it is essential that a relay with mains voltage contacts is fitted. These are available for currents up to several amperes, at 250v AC.

Construction

Figure 40 shows wired assembly on plain perforated board. The LDR is fitted vertically at one end of the board, soldered to pins inserted in the holes. Remember that the unused leads of IC1 should not touch each other or other connections.

The style of relay used will not affect working, provided it operates correctly with the current and voltage available. If necessary, this can be checked by placing a meter in series with the relay coil, which will show the change in current between light and dark conditions for the LDR. A very low resistance relay is not likely to be satisfactory, as it may require a heavy current. Should a relay of low resistance, but sufficiently sensitive to use, be to hand, a resistor may be placed in the collector circuit, to make up a total of about

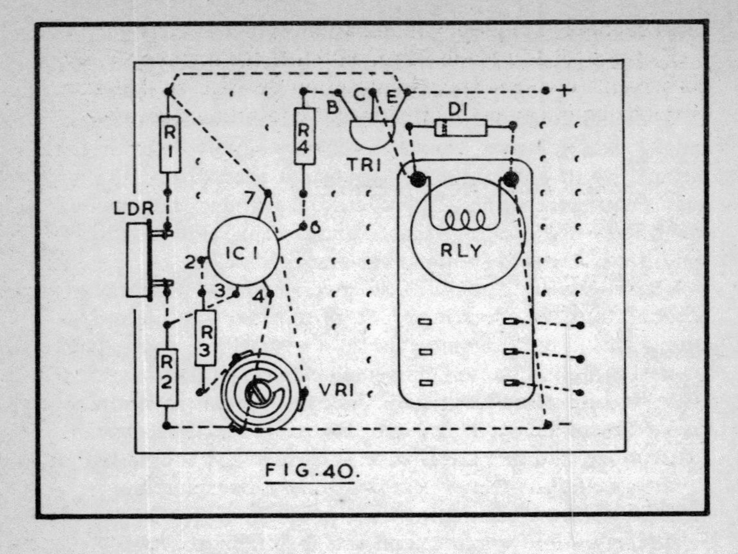

FIG.40.

150 ohm. Some high resistance relays, intended to operate with a very low current, will be satisfactory.

As the unit will probably be left operating for long periods, power is best drawn from a mains PSU. It would be in order to accommodate a small transformer and the other items in the case, for this purpose.

Bring leads from the relay contacts to sockets or terminals, or to a mains connector for high voltage. Usually, those two contacts which complete the circuit when the relay is energised will be in use.

A hole should be arranged in the case so that when the board is fitted in place, the LDR comes behind it. When the unit is to switch on a light in the same room, position the lamp or case so that light does not fall on the LDR, or a feedback effect will take place. This may be arranged by positioning the case with the LDR facing a window, and having a bedside or other lamp at some distance away in the room. The setting of VR1 considerably modifies the dark and light on-off switching, so has to be set for the condition under which it will be used. The setting is not critical, except for very small changes in low-level illumination.

For control of a mains lamp, with internal mains PSU, it is convenient to wire a 3-socket outlet on the case, so that any portable lamp may be plugged in, for control by the unit.

Components:- Dark Trigger (Figure 39)

Semiconductors

IC1	741
Tr1	2N3702
D1	1N4002

Resistors

R1	100k	R2	100k
R3	47k	R4	4.7k
VR1	2.5k pre-set		
LDR	ORP12 or similar		

Miscellaneous

Board, Case, etc.
RLY see text for relay details.

Light Relay

Operation of this circuit is the reverse of that described earlier. The relay is closed when light falls upon the light dependent resistor LDR, Figure 41. With this circuit, the fall in resistance of the LDR moves 3 of IC1 negative, and so output 6 furnishes base bias for PNP transistor Tr1. The rise in collector current then energises the relay.

If wished, the second set of relay contacts can be used to latch on the relay. To arrange this, connect one set of closing contacts from Tr1 collector to positive line, with a push-switch (normally closed) and limiting resistor of about 47 ohm in series. Then once the relay has been energised, current will continue to flow through these items to the relay winding, so that the relay remains closed until the push-switch is momentarily opened.

One use for the light relay is for garage lighting, where it switches on a lamp when car lamps illuminate the LDR. With

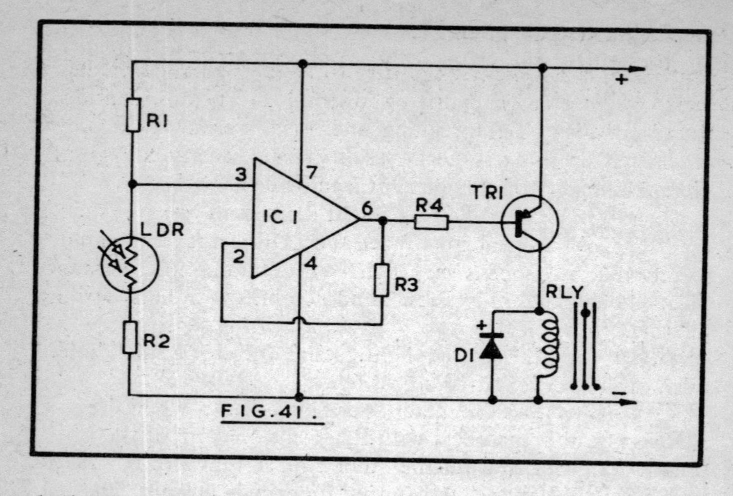

the latching circuit described, the garage light or a path light can remain on, until switched off.

For a board layout, and details of the relay, reference should be made to the earlier circuit. Relay current is about 50mA to 55mA with a 9v supply.

Components:- Light Relay (Figure 41)

Semiconductors

IC1	741
Tr1	2N3702
D1	1N4002

Resistors

R1	100k	R2	100k
R3	100k	R4	4.7k
LDR	ORP12 or similar		

Miscellaneous

Board, Case, Relay, etc.

Sensitive Enlarging Meter

This meter is sufficiently sensitive for very small areas of the image on the enlarger frame to be sampled. The illumination found provides a meter reading, and this is translated into exposures. In this way negatives of various density, and given different degrees of enlargement, can be dealt with.

Operation is controlled by a light dependent resistor, such as the ORP12, or equivalents. When the LDR is not illuminated, its internal resistance is very high. With illumination, resistance falls, and becomes only a few hundred ohms with fairly strong lighting. In this instrument, only a part of the sensitive surface of the LDR is used, light passing through a small hole in an opaque cover fitted across it.

Figure 42 shows the meter operating circuit. When the LDR is not illuminated, it with R2 forms one arm of the potential divider, and R1 the other arm, so that 3 of IC1 takes up a certain potential. IC1 output 6 depends on this. The output transistors in IC1 form one arm of a bridge, and VR1 the other arm. VR1 is set so that balance arises, and the meter M shows no current, because the potential at 6 of IC1 and the wiper of VR1 is the same.

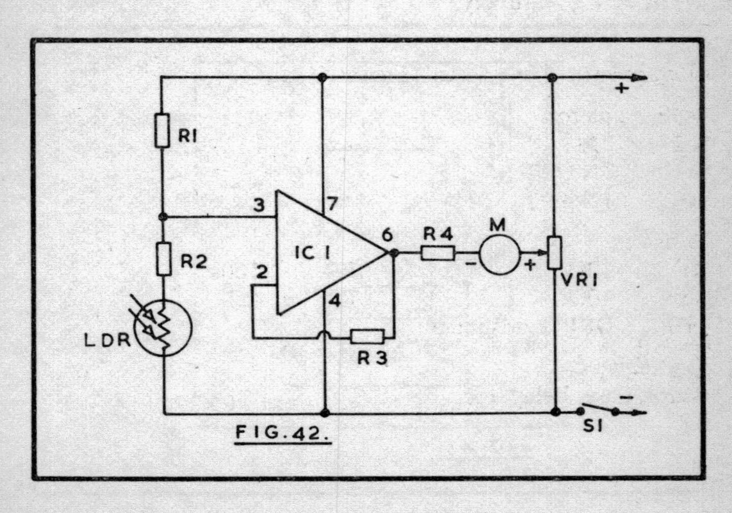

FIG. 42.

With illumination of the LDR, its resistance falls, and so the voltage at 3 drops. Point 6 follows this, producing a reading on the meter. R4 is to limit current.

Meter M is 1mA full-scale. When VR1 has been set so that no reading is obtained with the LDR covered, even moderate light will produce a full-scale deflection. It is thus necessary to cover the LDR with an opaque card in which a small hole has been punched, so that sensitivity is reduced. This has the advantage that small areas of the image can be checked, and the meter then takes up a reading which corresponds to their brightness. Operation is from a 9v battery.

Construction

Figure 43 shows the small components assembled on a perforated board. Leads 1, 5 and 8 of IC1 are unused. Underneath numbering of the IC is as shown earlier.

It is necessary that the LDR will be near the plane of the enlarger paper holder. For this reason, and to allow easy checking of individual areas of the image, it is fitted in a small holder, with thin flexible leads to R2 and the negative line on the board. The holder can be made from wood, with a

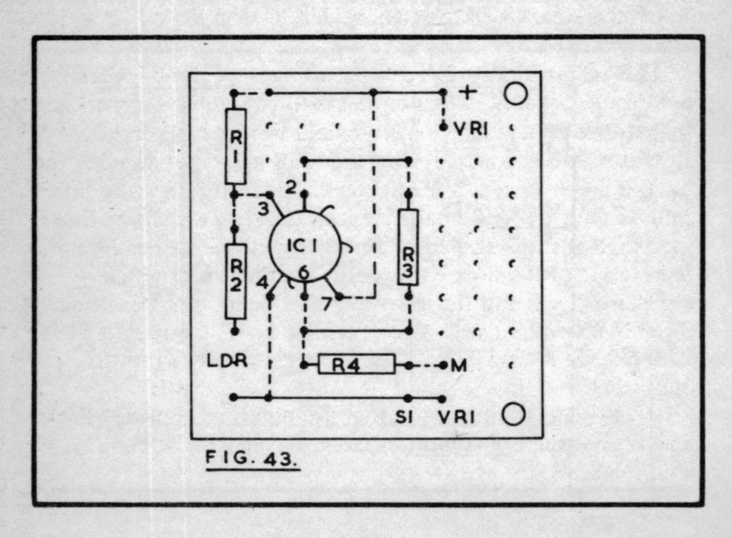

FIG. 43.

depression cut from it to take the LDR, or may be any very shallow container or box of small size. The total depth needed is about ½in or 12mm. Fit a handle, such as a disused felt pen, and take the leads through this.

The board, battery, VR1, S1 and meter are fitted in a case, and one with a sloping front is most suitable. Dimensions will depend on the meter, but about 4 x 4 x 4in or 102mm cube is suitable. Fit VR1 below the meter, with pointer knob and scale. Connect VR1 outer tags to positive and negative lines in such a way that the meter reading rises when VR1 is turned clockwise. The indicator box is placed a little to one side off the enlarger paper holder.

The LDR is covered with an opaque card, in which a hole about 1/8in in diameter (3mm) or a little smaller is made. As the best size for this hole depends to some extent on the enlarger, experiment as necessary before fixing the aperture. An enlarger with condenser illumination from a high intensity lamp will make necessary a smaller hole than one using a normal power lamp and diffused illumination. VR1 is set so that the meter gives no reading with the LDR covered. The opening above the LDR is of such a size that the meter reading does not go off the scale when the enlarger lens is stopped down to a usual f-number.

There are various ways of calibrating the enlarging meter in terms of exposures. The simplest is to have a pre-set mark on M, such as the central scale position, and to adjust the enlarger stop until this is reached, with the LDR aperture sampling the lightest area which is to print black. This automatically takes care of all degrees of enlargement or negative density within the range of stops available. Another method is to calibrate M in terms of the time, in seconds, required to produce complete black, and this gives more scope in use. Development should be standardised, as for example at 1½ minutes at 20° C, with one grade of paper and developer, so that results are duplicated.

It will generally be found that illumination over the whole area is unequal, but this arises from the enlarger design.

Components:- Sensitive Enlarging Meter (Figure 42)

Semiconductors
IC1 741

Resistors

R1	100k	R2	100k
R3	100k	R4	1k
VR1	2.5k Lin. potentiometer		
LDR	ORP12 or similar		

Miscellaneous
M 1mA moving coil meter
Board, Case, Switch, etc.

Sensitive Sound Relay

This unit closes an external circuit when actuated by a sound which is above a pre-set level. There are a number of uses for it in the home and elsewhere. It can give warning of the operation of a doorbell or telephone bell, without any actual connection to the bell or phone, and operate a buzzer, bell or lamp at a distance. This is useful when, as example, activities in the kitchen or elsewhere, or watching TV, may result in the phone going unheard. To avoid this, a second bell, operated by the unit, can be fitted where it will not go unnoticed. A similar arrangement can be used for a doorbell. With the deaf, or a person hard of hearing, the unit can switch on an indicator lamp, prominently situated, and this will show that the door-bell has been rung.

The circuit is so arranged that its operation of the warning bell or indicator lamp can be intermittent, and will cease when the sound stops. Or it can be switched to latch on, and its bell or lamp will then remain in action, once triggered, until the unit is switched off by hand.

The circuit is shown in Figure 44. M is a small crystal microphone insert, used to pick up the external sound. It is possible to use a miniature loudspeaker here, but it must be of quite high impedance (75–80 ohm) and sensitivity is reduced. With the speaker, an isolating capacitor of about 10μF must be included in the circuit, to avoid shorting R2.

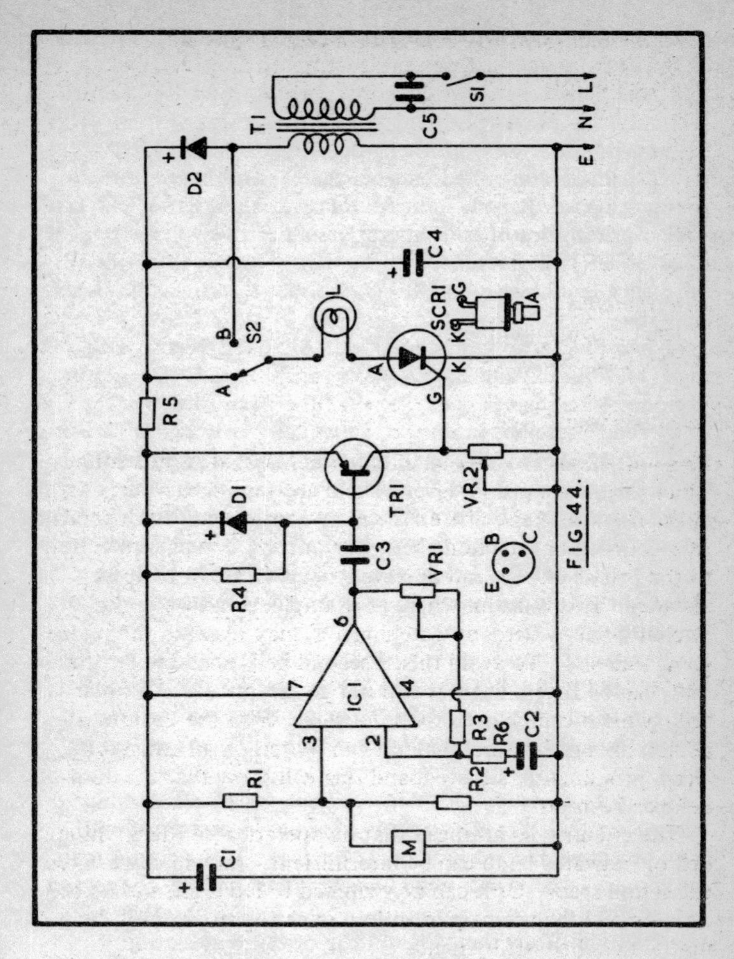

FIG. 44.

IC1 is the audio amplifier, to raise the signal level of sounds reaching the microphone M. VR1 is a pre-set gain control. The maximum gain available normally has to be reduced, or the circuit will be operated by quite small external sounds. The amplifier audio is available at C3.

Tr1 is a PNP transistor, with base held positive by R4. In

these circumstances, collector current through VR2 is negligible. When audio signals are present, rectification by D1 makes Tr1 base negative so that collector current flows. As a result, Tr1 collector moves positive, due to the voltage drop in VR2. C1 and R5 are to smooth the supply to IC1 and Tr1.

The silicon controlled rectifier SCR1 is normally not conducting, so that the indicator lamp L is not lit. When Tr1 collector moves positive, this applies a positive voltage to gate G of SCR1, so it goes into conduction. Current then flows through the indicator lamp, showing that the circuit has been actuated.

When the 2-way switch (S2) is at A, SCR1 receives DC, from rectifier D2 and the reservoir capacitor C4. As a result, conduction is maintained. So when the circuit has been triggered, L remains lit until it is manually switched off.

With the switch (S2) at B, the SCR receives AC directly from the secondary of T1. Since the anode voltage drops to zero with each cycle, the SCR ceases to conduct, but is returned to the conducting condition so long as gate G is positive. In these circumstances, L flashes only so long as the bell rings. When the operating sound stops, the circuit returns to the off condition.

T1 is a bell transformer with 8v secondary, and D2 provides the necessary supply for earlier stages, so that the whole unit can run from AC mains. C5 is to help suppress interference, carried into the unit by the mains. A 9v transformer can be used.

Lamp L can be on an extension lead, for visual warning elsewhere. A 12v 3 watt bulb is suitable. A buzzer or bell may be connected to these points instead. If the circuit is to latch on, a resistor of a few hundred ohms must be connected across this item, as interruption of SCR1 anode by the trembler will allow the SCR to drop out of conduction.

The SCR can also control a relay, whose contacts close to switch an external circuit. This is most suitable when complete isolation is required between unit and bell or signal lamp, and will be used for a mains voltage indicator lamp.

The rating of T1 (and D2) will depend largely on what load is to be operated by the SCR. For normal use, current taken by IC1 and Tr1 can be ignored. Should SCR1 be intended to control a relay or low-current bulb, T1 need be rated at only 100mA.

Construction

Figure 45 shows a board layout. The microphone M will have one case connection, and this must be taken to the negative or earth line, using a short screened lead. The microphone can

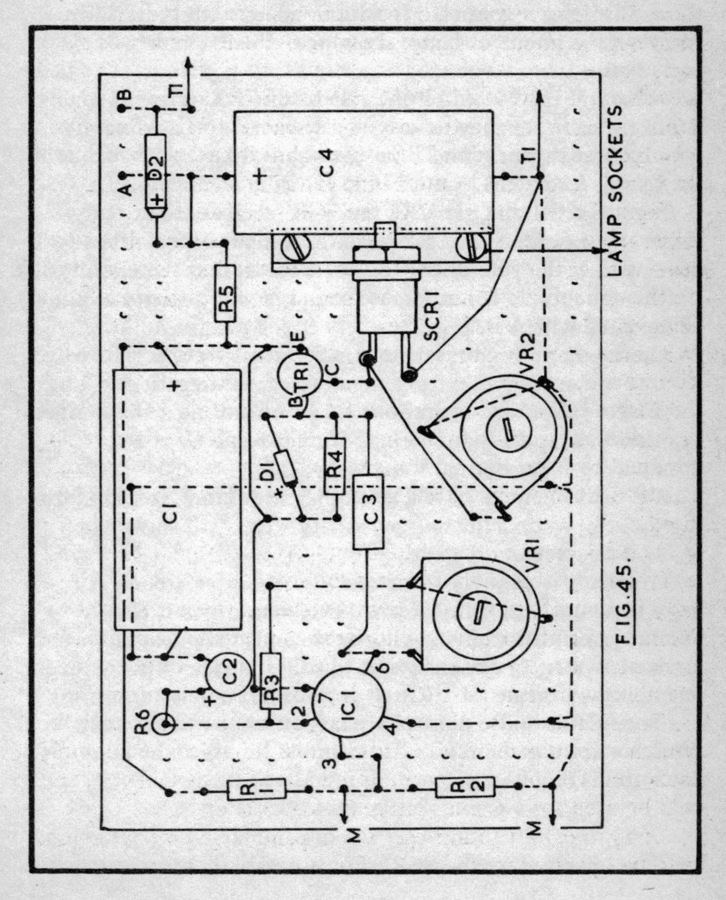

FIG.45.

then be fixed behind an aperture in the case.

Details on the use of boards in this way have been provided earlier. SCR1 is mounted on a metal bracket, which also provides the anode connection. Carefully check throughout for omitted leads, or possible shorts or other errors. Flexible leads are provided at points T1, to take to the transformer secondary.

A metal box is most suitable, as this provides screening though it is not essential. Mount the board with bolts with spacers or extra nuts, to give clearance. Fit S1, and two sockets for the low voltage lamp circuit, to the front. The box is earthed. T1 is bolted to one side of the box. Provide mains connections in the way described elsewhere, to assure safety.

Take leads from A and B on the board, to a 2-way switch on the front, marking A "Latch" and B "Intermittent".

For an initial test, set VR1 and VR2 each at about half value. VR1 controls the gain of IC1, and VR2 the voltage developed at the gate of SCR1. Set VR2 so that the circuit triggers reliably with a moderate sound, and sensitivity can then be adjusted with VR1.

There is considerable latitude in settings. Very sensitive settings are not suitable for general usage, as triggering will be too easy. To localise operation, place the unit near the phone or other bell, and adjust working so that response is not obtained to other normal sounds.

It is convenient to have a small indicator lamp, to plug into the sockets, when adjusting sensitivity. This will show when SCR1 is triggered.

Triggering is possibly from heavy mains interference. C5 helps suppress this. The screened box mentioned is also helpful. In addition, avoid placing the unit very near to domestic wiring or appliances, if this fault arises. Also reduce sensitivity by means of VR2. It is possible to delay triggering, so that isolated static pulses have less effect, by placing a capacitor from gate of SCR1 to cathode K. It can be about 20nF to 0.1μF.

Components:- Sensitive Sound Relay (Figure 44)

Semiconductors

IC1	741		
Tr1	AC128		
SCR1	50v 3A type		
D1	OA47	D2	50/100v 2A type

Resistors

R1	100k	R2	100k
R3	10k	R4	470k
R5	2.2k	R6	1k
VR1	250k pre-set	VR2	1k pre-set

Capacitors

C1	1000μF 15v	C2	10μF 10v
C3	22nF	C4	2200μF 15v
C5	10nF 600v		

Miscellaneous

T1 250v/8v 1A transformer
M Crystal microphone insert
L 12v 3W bulb
S1 Mains toggle switch
S2 2-way switch
Board, Case, Sockets, etc.

Sensitive Touch Switch

This switch is of the type operated by touching a surface, and in this form it can be used to bring on a low voltage bedside light. It can also be used as a rain indicator. For this method of operation, spots of rain on the touch plate operate the circuit, which can light a warning lamp.

In Figure 46 input 3 of IC1 is held by resistors R1 and R2. Inverting input 2 is taken to the slider of VR1, which allows the threshold of operation to be set. When 2 moves negative, due to a touch on the plate or other reduced resistance, this effect is amplified, and 6 moves positive.

The silicon controlled rectifier SCR1 is normally not conducting so the lamp is extinguished. When its gate is made

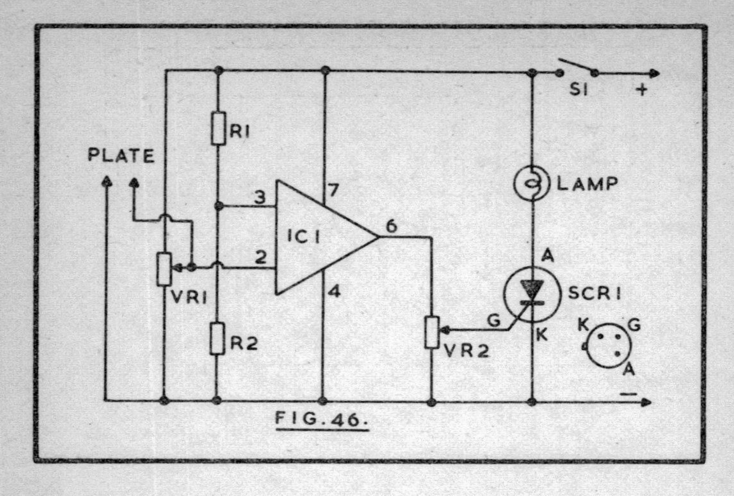

FIG.46.

positive, it conducts, and remains in this condition until the supply is broken by opening the switch S1. VR2 allows the sensitivity of operation to be set to suit the SCR and purpose in view.

Current can be drawn from a 9v battery supply. The easiest way to set up the circuit is to clip a high resistance voltmeter from 6 to the negative line (the latter is negative). Set VR1 so that the voltage is low, but rises sharply when the circuit from 2 is touched. VR2 is then set with the wiper near negative, and is slowly rotated until the rise in voltage at 6 triggers the SCR. Lack of a sharp voltage change at 6 would arise from VR1 being wrongly set, while failure of the SCR to trigger, or the lamp lighting at once when S1 is closed, would be caused by wrong adjustment of VR2.

Construction

The touch plate is made by using a piece of circuit board with six or eight strips an inch or so long, and joining together strips 1, 3 and 5 for one side of the circuit, and 2, 4 and 6 for the other side. This plate is fixed to the top of the case with adhesive or screws (clear away adjacent foils) or can be on an extension lead to place outside a window.

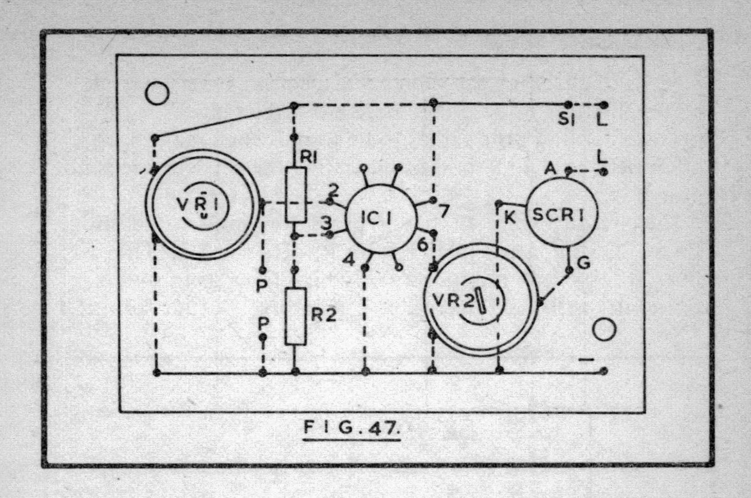

FIG.47.

Figure 47 shows a board layout. Points P–P are connected to the touch plate, and L–L to the lampholder. S1 connects to the positive line. The unused leads of IC1 mush not touch each other, or any other circuits.

Components:- Sensitive Touch Switch (Figure 46)

Semiconductors

IC1	741		
SCR1	50v 1A type		

Resistors

R1	100k	R2	100k
VR1	500k Lin. potentiometer		
VR2	2.5k Lin. potentiometer		

Miscellaneous

Lamp 12v 0.3A or as convenient
S1 On-off switch
Board, Case, Lampholder, etc.

IC Capacitance Bridge

A means of checking the values of capacitors is very useful, and the instrument described here has three ranges for this purpose. It allows the values to be found when markings are obliterated, or the method of indicating the capacitance is not known.

The circuit, Figure 48, has an oscillator, and sensing bridge. The oscillator consists of IC1, with R1, R2 and C1. The values are fixed, to produce a fairly high tone, as this gives best results with the lower value capacitors. Output from IC1

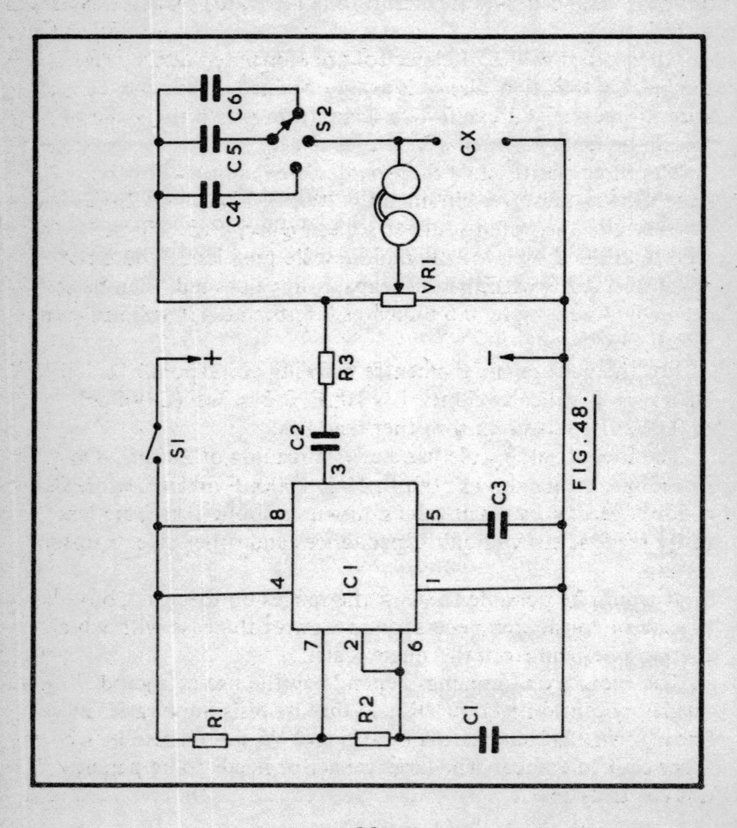

FIG. 48.

is from 3, and is coupled to the bridge by C2. This part of the circuit runs from a 9v or other convenient supply, and has the on-off switch S1.

The bridge consists of upper and lower portions of the linear potentiometer VR1 for one side, and one of the capacitors C4, C5 or C6, with the unknown capacitor Cx for the other side. It will be seen that where the capacitor selected by S2 equals the value of the capacitor connected at Cx, and VR1 wiper is at the centre of the track, equal potentials will arise each side of the phones. In these circumstances, the bridge is balanced, and the audio tone generated by IC1 ceases to be heard.

When other values of capacitor are connected at Cx, VR1 can be re-set, so that blance is again obtained. Thus the potentiometer VR1 can have a scale, from which the value of Cx can be read.

The three positions of S2 provide three ranges. For the scale shown, these are obtained by 200pF, 20nF, and 2μF. Referring to the potentiometer scale, Figure 49, it will be seen that 2 is shown for the middle scale position. Where the range selected is with the 2μF capacitor, values will then be as marked. As example, 0.5 indicates 0.5μF, and 4.7 will indicate 4.7μF, and so on.

For the next range, the centre marking corresponds to 20nF. So 5 indicates 50nF, 1 is 10nF, 0.2 is 2nF (2,000pF or 0.002μF) and so on for other readings.

The lowest value scale has a centre reading of 200pF. On this range, 5 thus shows 500pF, 10 is 1000pF or 1nF, while 0.5 is 50pF, etc. Very small values, down to about 10pF, are less easily read, as stray circuit capacitances and other effects upset results.

It would be possible to mark the ranges on the scale, but this is not considered necessary, in view of the ease with which they can be read from the single scale.

The accuracy of readings depends on the use of a good quality potentiometer at VR1, so that its resistance varies in linear form. The capacitors C4, C5 and C6 should also be of 1 per-cent tolerance. The large capacitor needs to be paper, not electrolytic.

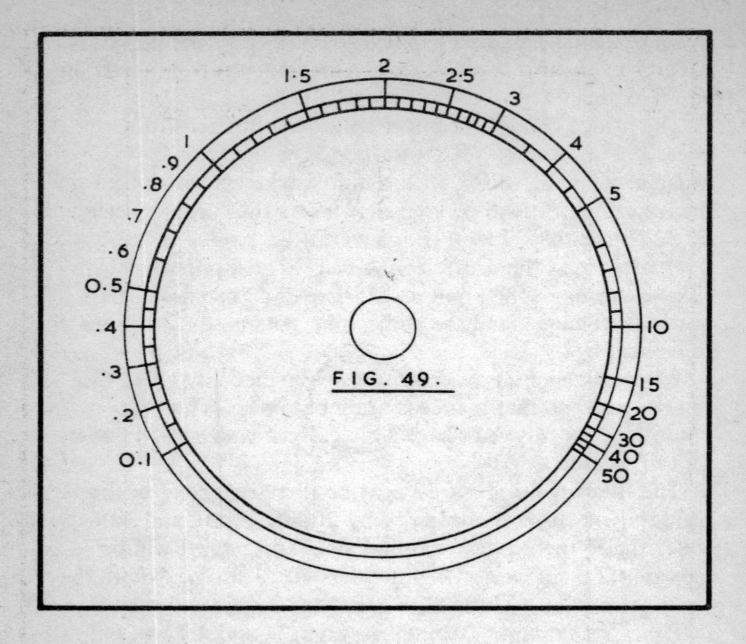

FIG. 49.

Listening for the null or zero position is best with 2k or similar headphones. The setting is sharp with capacitors of good quality at Cx, but a wide setting, or failure to obtain any reasonable indication of value, indicates leakage in the capacitor.

Construction

It is convenient to assemble IC1, R1, R2, R3, C1, C2 and C3 on a board using the layout shown for the code oscillator. This part of the capacitance meter can be tested by checking that an audio tone is obtained at R3.

A case about 6 x 4in and 2in deep (152 x 102 x 51mm) is suitable for the whole assembly. Mount VR1 centrally on the front, and fit the scale shown, using a pointer knob, or knob with cursor as in Figure 50. Fit S2 to the right of VR1. C4, C5 and C6 are connected directly between the tags of S2, and VR1. Also fit a 3.5mm jack socket for the headphone plug,

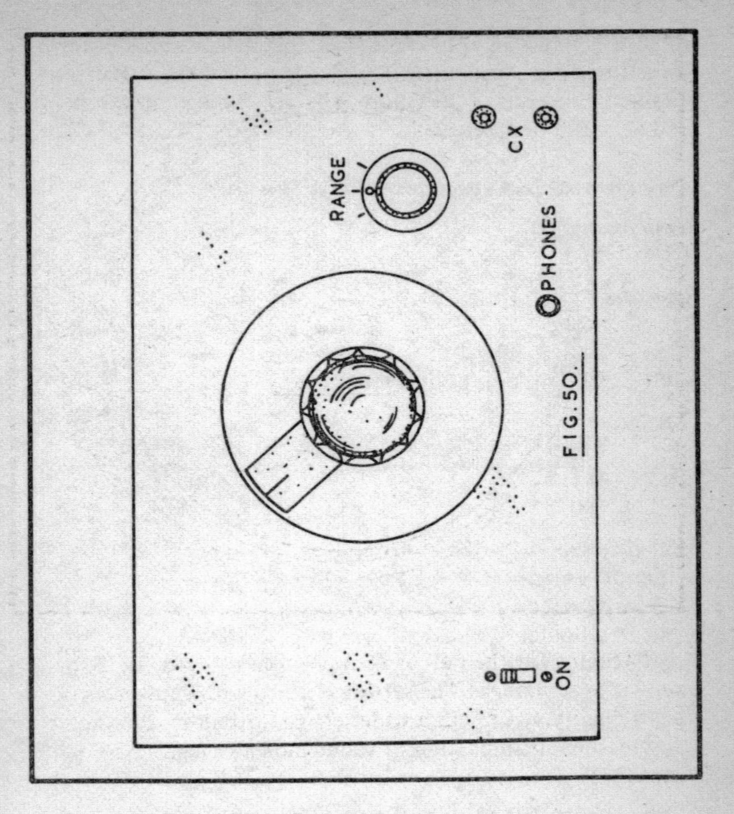

FIG. 50.

and two insulated terminals for the unknown capacitor. The IC board can then be mounted with two small bolts, and the negative line and tone circuit from R3 can be connected.

The switch positions are marked with the middle-scale settings for VR1. That is, 200pF, 20nF or 0.02μF, and 2μF. If a spare 1 per-cent capacitor is to hand, such as 20nF, adjust VR1 for the null described, then fix the knob with its pointer indicating 2 on the scale. Two short flexible leads, with small clips, will prove of use when checking capacitors.

General purpose multirange meters usually have one or more resistance ranges, but if wished resistance ranges can be added to the bridge. For three resistance ranges (in addition to the three for capacitances) S2 needs to be 6-way. Ranges obtained by means of 200 ohm, 20k, and 2 megohm can be used (2 x 100 ohm, etc.).

Components:- IC Capacitance Bridge (Figure 48)

Semiconductors

IC1	555		

Resistors

R1	220k	R2	2.2k
R3	680Ω		
VR1	5k Lin. potentiometer		

Capacitors

C1	10nF	C2	0.1μF
C3	0.1μF	C4	200pF 1%
C5	0.02μF 1%	C6	2μF 1%

Miscellaneous

S1 On-off switch
S2 3-way switch
8-pin DIL holder
Board, Case, etc.

Notes

Notes

Notes

Notes

Notes

OTHER BOOKS RECOMMENDED FOR BEGINNERS

227: BEGINNERS GUIDE TO BUILDING ELECTRONIC PROJECTS **price £1.50**

Author: R. A. Penfold

ISBN 0 900162 68 6 1977

Approx. size: 180 × 108 mm 112 pages

The purpose of this book is to enable the complete beginner to tackle the practical side of electronics, so that he or she can confidently build the electronic projects that are regularly featured in the popular magazines and books.

Subjects such as component identification, tools, soldering, various constructional methods (Matrixboard, Veroboard, P.C.B.), cases, legends, etc. are covered in detail and practical examples in the form of simple projects are given.

228: ESSENTIAL THEORY FOR THE ELECTRONICS HOBBYIST **price £1.25**

Author: G. T. Rubaroe, T.Eng.(C.E.I.), Assoc.I.E.R.E.

ISBN 0 900162 69 4 1977

Approx. size: 180 × 108 mm 128 pages

In any hobby activity a background knowledge of the subject can considerably increase the enjoyment and satisfaction one derives from it. This point of view applies, without any reservations whatsoever, to electronics.

The object of this book is to supply the hobbyist with a background knowledge tailored to meet his or her specific requirements and the author has brought together the relevant material and presented it in a readable manner with minimum recourse to mathematics.

Many formulae having a practical bearing are presented in this book and purpose-designed examples are employed to illustrate their applications.

BP48: ELECTRONIC PROJECTS FOR BEGINNERS **price £1.35**

Author: F. G. Rayer, T.Eng.(C.E.I.), Assoc.I.E.R.E.

ISBN: 0 85934 054 6 1978

Approx. size: 180 × 108 mm 128 pages

Another book written by the very experienced author – Mr F. G. Rayer – and in it the newcomer to electronics, will find a wide range of easily-made projects. Also, there are a considerable number of actual component and wiring layouts, to aid the beginner.

Furthermore, a number of projects have been arranged so that they can be constructed without any need for soldering and, thus, avoid the need for a soldering iron.

Also, many of the later projects can be built along the lines of those in the "No Soldering" section so this may considerably increase the scope of projects which the newcomer can build and use.

ELEMENTS OF ELECTRONICS – an on-going series **price (each volume) £2.25**

BP62: BOOK 1, – The Simple Electronic Circuit and Components

ISBN: 0 900162 82 1 1979

BP63: BOOK 2 – Alternating Current Theory

ISBN: 0 900162 83 X 1979

BP64: BOOK 3 – Semiconductor Technology

ISBN: 0 900162 84 8 1979

Author: F. A. Wilson, C.G.I.A., C.Eng., F.I.E.E., F.B.I.M.

Approx. sizes: 180 × 108 mm each 224 pages

The aim of this series of books can be stated quite simply – it is to provide an inexpensive introduction to modern electronics so that the reader will start on the right road by thoroughly understanding the fundamental principles involved.

That the books are inexpensive is perhaps obvious. By accepting their modest size the total information is that which in normal size text books would cost some 2–3 times as much.

Although written especially for readers with no more than ordinary arithmetical skills, the use of mathematics is not avoided, and all the mathematics required is taught as the reader progresses. There is little doubt that some skill in electronics and mathematics will be an asset to almost everybody in the future, destined as we are to be controlled more and more by computers, microprocessors and the like. There is also much to offer those who need revision or to be brought up to date with fundamentals, those with grown-up children asking awkward questions or preparing for entry to a college or even those who may find a different approach refreshing.

The course is not written to a specific syllabus but concentrates on the understanding of the important concepts central to electronics rather than continually digressing over the whole field on the basis that once the fundamentals are mastered, the technicalities of most other things are soon revealed. The author anticipates where the difficulties lie and hopefully guides the reader through them.

Each book is a complete treatise of a particular branch of the subject and, therefore, can be used on its own with one proviso, that the later books do not duplicate material from their predecessors, thus a working knowledge of the subjects covered by the earlier books is assumed.

BOOK 1: This book contains all the fundamental theory necessary to lead a full understanding of the simple electronic circuit and its main components.

BOOK 2: This book continues with alternating current theory without which there can be no comprehension of speech, music, radio, television or even the electricity mains.

BOOK 3: Follows on semiconductor technology, leading up to transistors and integrated circuits.

Books 1–3 therefore constitute a complete but inexpensive basic electronic theory course of inestimable help in either career or hobby. Other books following are on specialised electronic subjects, they are all built on this earlier groundwork.

Two further books are available:

BP77: BOOK 4 – Microprocessing Systems and Circuits
BP89: BOOK 5 – Communication

These two volumes are priced at £2.95 each.

Please note overleaf is a list of other titles that are available in our range of Radio, Electronics and Computer Books.

These should be available from all good Booksellers, Radio Component Dealers and Mail Order Companies.

However, should you experience difficulty in obtaining any title in your area, then please write directly to the publisher enclosing payment to cover the cost of the book plus adequate postage.

If you would like a complete catalogue of our entire range of Radio, Electronics and Computer Books then please send a Stamped Addressed Envelope to:

BERNARD BABANI (publishing) LTD
THE GRAMPIANS
SHEPHERDS BUSH ROAD
LONDON W6 7NF
ENGLAND

160	Coil Design and Construction Manual	1.50p
202	Handbook of Integrated Circuits (IC's) Equivalents & Substitutes	1.75p
205	First Book of Hi-Fi Loudspeaker Enclosures	95p
221	28 Tested Transistor Projects	1.25p
222	Solid State Short Wave Receivers for Beginners	1.25p
223	50 Projects Using IC CA3130	1.25p
224	50 CMOS IC Projects	1.25p
225	A Practical Introduction to Digital IC's	1.25p
226	How to Build Advanced Short Wave Receivers	1.20p
227	Beginners Guide to Building Electronic Projects	1.50p
228	Essential Theory for the Electronics Hobbyist	1.25p
RCC	Resistor Colour Code Disc	20p
BP1	First Book of Transistor Equivalents and Substitutes	95p
BP2	Handbook of Radio, TV & Ind. & Transmitting Tube & Valve Equiv.	60p
BP6	Engineers and Machinists Reference Tables	75p
BP7	Radio and Electronic Colour Codes and Data Chart	40p
BP14	Second Book of Transistor Equivalents and Substitutes	1.75p
BP24	52 Projects Using IC741	1.25p
BP27	Chart of Radio Electronic Semiconductor and Logic Symbols	50p
BP32	How to Build Your Own Metal and Treasure Locators	1.75p
BP33	Electronic Calculator Users Handbook	1.50p
BP34	Practical Repair and Renovation of Colour TVs	1.25p
BP35	Handbook of IC Audio Preamplifier & Power Amplifier Construction	1.75p
BP36	50 Circuits Using Germanium, Silicon and Zener Diodes	95p
BP37	50 Projects Using Relays, SCR's and TRIACs	1.75p
BP39	50 (FET) Field Effect Transistor Projects	1.75p
BP40	Digital IC Equivalents and Pin Connections	2.50p
BP41	Linear IC Equivalents and Pin Connections	3.75p
BP42	50 Simple L.E.D. Circuits	95p
BP43	How to Make Walkie-Talkies	1.50p
BP44	IC555 Projects	1.95p
BP45	Projects in Opto-Electronics	1.75p
BP46	Radio Circuits Using IC's	1.35p
BP47	Mobile Discotheque Handbook	1.35p
BP48	Electronic Projects for Beginners	1.35p
BP49	Popular Electronic Projects	1.45p
BP50	IC LM3900 Projects	1.35p
BP51	Electronic Music and Creative Tape Recording	1.75p
BP52	Long Distance Television Reception (TV–DX) for the Enthusiast	1.95p
BP53	Practical Electronic Calculations and Formulae	2.95p
BP55	Radio Stations Guide	1.75p
BP56	Electronic Security Devices	1.75p
BP57	How to Build Your Own Solid State Oscilloscope	1.50p
BP58	50 Circuits Using 7400 Series IC's	1.50p
BP59	Second Book of CMOS IC Projects	1.50p
BP60	Practical Construction of Pre-amps, Tone Controls, Filters & Attn	1.45p
BP61	Beginners Guide to Digital Techniques	95p
BP62	Elements of Electronics – Book 1	2.25p
BP63	Elements of Electronics – Book 2	2.25p
BP64	Elements of Electronics – Book 3	2.25p
BP65	Single IC Projects	1.50p
BP66	Beginners Guide to Microprocessors and Computing	1.75p
BP67	Counter Driver and Numeral Display Projects	1.75p
BP68	Choosing and Using Your Hi-Fi	1.65p
BP69	Electronic Games	1.75p
BP70	Transistor Radio Fault-Finding Chart	50p
BP71	Electronic Household Projects	1.75p
BP72	A Microprocessor Primer	1.75p
BP73	Remote Control Projects	1.95p
BP74	Electronic Music Projects	1.75p
BP75	Electronic Test Equipment Construction	1.75p
BP76	Power Supply Projects	1.75p
BP77	Elements of Electronics – Book 4	2.95p
BP78	Practical Computer Experiments	1.75p
BP79	Radio Control for Beginners	1.75p
BP80	Popular Electronic Circuits – Book 1	1.95p
BP81	Electronic Synthesiser Projects	1.75p
BP82	Electronic Projects Using Solar-Cells	1.95p
BP83	VMOS Projects	1.95p
BP84	Digital IC Projects	1.95p
BP85	International Transistor Equivalents Guide	2.95p
BP86	An Introduction to Basic Programming Techniques	1.95p
BP87	Simple L.E.D. Circuits – Book 2	1.35p
BP88	How to Use Op-Amps	2.25p
BP89	Elements of Electronics – Book 5	2.95p
BP90	Audio Projects	1.95p
BP91	An Introduction to Radio DX-ing	1.95p
BP92	Electronics Simplified – Crystal Set Construction	1.95p
BP93	Electronic Timer Projects	1.95p
BP94	Electronic Projects for Cars and Boats	1.95p
BP95	Model Railway Projects	1.95p
BP96	C B Projects	1.95p
BP97	IC Projects for Beginners	1.95p
BP98	Popular Electronic Circuits – Book 2	2.25p
BP99	Mini-Matrix Board Projects	1.95p
BP100	An Introduction to Video	1.95p
BP101	How to Identify Unmarked IC's	65p
BP102	The 6809 Companion	1.95p
BP103	Multi-Circuit Board Projects	1.95p
BP104	Electronic Science Projects	2.25p
BP105	Aerial Projects	1.95p
BP106	Modern Op-Amp Projects	1.95p